ACROSS THE FRONTIERS

the text of this book is printed on 100% recycled paper

WORLD PERSPECTIVES

Volumes already published

I	Approaches to God	*Jacques Maritain*
II	Accent on Form	*Lancelot Law Whyte*
III	Scope of Total Architecture	*Walter Gropius*
IV	Recovery of Faith	*Sarvepalli Radhakrishnan*
V	World Indivisible	*Konrad Adenauer*
VI	Society and Knowledge	*V. Gordon Childe*
VII	The Transformations of Man	*Lewis Mumford*
VIII	Man and Materialism	*Fred Hoyle*
IX	The Art of Loving	*Erich Fromm*
X	Dynamics of Faith	*Paul Tillich*
XI	Matter, Mind and Man	*Edmund W. Sinnott*
XII	Mysticism: Christian and Buddhist	*Daisetz Teitaro Suzuki*
XIII	Man's Western Quest	*Denis de Rougemont*
XIV	American Humanism	*Howard Mumford Jones*
XV	The Meeting of Love and Knowledge	*Martin C. D'Arcy, S.J.*
XVI	Rich Lands and Poor	*Gunnar Myrdal*
XVII	Hinduism: Its Meaning for the Liberation of the Spirit	*Swami Nikhilananda*
XVIII	Can People Learn to Learn?	*Brock Chisholm*
XIX	Physics and Philosophy	*Werner Heisenberg*
XX	Art and Reality	*Joyce Cary*
XXI	Sigmund Freud's Mission	*Erich Fromm*
XXII	Mirage of Health	*René Dubos*
XXIII	Issues of Freedom	*Herbert J. Muller*
XXIV	Humanism	*Moses Hadas*
XXV	Life: Its Dimensions and Its Bounds	*Robert M. MacIver*

WORLD PERSPECTIVES • *Volume Forty-eight*

Planned and Edited by **RUTH NANDA ANSHEN**

ACROSS THE FRONTIERS

WERNER HEISENBERG

Translated from the German by Peter Heath

HARPER TORCHBOOKS
Harper & Row, Publishers
New York, Evanston, San Francisco, London

First published in German as *Schritte über Grenzen*. Copyright © R. Piper & Co. Verlag, Munich, 1971. First publication of

1. "Das Naturbild und die Strukter der Materie"
2. "Die Bedeutung des Schonen in der exakten Naturwissenschaft"

by Chr. Belser Verlag, Stuttgart, 1967.

 For information address Harper & Row, Publishers, Inc., 10 East 53rd Street, New York, N.Y. 10022. Published simultaneously in Canada by Fitzhenry & Whiteside Limited, Toronto.

First HARPER TORCHBOOK edition published 1975

STANDARD BOOK NUMBER: 06–131859–0

75 76 77 78 79 10 9 8 7 6 5 4 3 2 1

Contents

By the same author
in
WORLD PERSPECTIVES

Physics and Philosophy
Physics and Beyond

World Perspectives

What This Series Means

It is the thesis of *World Perspectives* that man is in the process of developing a new consciousness which, in spite of his apparent spiritual and moral captivity, can eventually lift the human race above and beyond the fear, ignorance, and isolation which beset it today. It is to this nascent consciousness, to this concept of man born out of a universe perceived through a fresh vision of reality, that *World Perspectives* is dedicated.

My Introduction to this Series is not of course to be construed as a prefatory essay for each individual book. These few pages simply attempt to set forth the general aim and purpose of the Series as a whole. They try to point to the principle of permanence within change and to define the essential nature of man, as presented by those scholars who have been invited to participate in this intellectual and spiritual movement.

Man has entered a new era of evolutionary history, one in which rapid change is a dominant consequence. He is contending with a fundamental change, since he has intervened in the evolutionary process. He must now better appreciate this fact and then develop the wisdom to direct the process toward his fulfillment rather than toward his destruction. As he learns to apply his understanding of the physical world for practical purposes, he is, in reality, extending his innate capacity and augmenting his ability and his need to communicate as well as his ability to think and to create. And as a result, he is substituting a goal-directed evolutionary process in his struggle against environmental hardship for the slow, but effective, biological evolution which produced modern man through mutation and natural selection. By intelligent intervention in the evolutionary process man has greatly accelerated and greatly expanded the range of his possibilities. But he has not changed the basic fact that it remains a trial and error process, with the danger of taking paths that lead to sterility of mind and heart, moral apathy and intellectual inertia; and even producing social dinosaurs unfit to live in an evolving world.

Only those spiritual and intellectual leaders of our epoch who have a paternity in this extension of man's horizons are invited to participate in the Series: those who are aware of the truth that beyond the divisiveness among men there exists a primordial unitive power since we are all bound together by a common humanity more fundamental than any unity of dogma; those who recognize that the centrifugal force which has scattered and atomized mankind must be replaced by an integrating structure and process capable of bestowing meaning and purpose on existence; those who realize that science itself, when not inhibited by the limitations of its own methodology, when chastened and humbled, commits man to an indeterminate range of yet undreamed consequences that may flow from it.

Virtually all of our disciplines have relied on conceptions which are now incompatible with the Cartesian axiom, and with the static world view we once derived from it. For underlying the new ideas, including those of modern physics, is a unifying order, but it is not causality; it is purpose, and not the purpose of the universe and of man, but the purpose *in* the universe and *in* man. In other words, we seem to inhabit a world of dynamic process and structure. Therefore we need a calculus of potentiality rather than one of probability, a dialectic of polarity, one in which unity and diversity are redefined as simultaneous and necessary poles of the same essence.

Our situation is new. No civilization has previously had to face the challenge of scientific specialization, and our response must be new. Thus this Series is committed to ensure that the spiritual and moral needs of man as a human being and the scientific and intellectual resources at his command for *life* may be brought into a productive, meaningful and creative harmony.

In a certain sense we may say that man now has regained his former geocentric position in the universe. For a picture of the Earth has been made available from distant space, from the lunar desert, and the sheer isolation of the Earth has become plain. This is as new and as powerful an idea in history as any that has ever been born in man's consciousness. We are all becoming seriously concerned with our natural environment. And this concern is not only the result of the warnings given by biologists, ecologists and conservationists. Rather it is the result of a deepening awareness that something new has happened, that the planet Earth is a unique and precious place. Indeed, it may not be a mere coincidence that this awareness should have been born at the exact moment when man took his first step into outer space.

This Series endeavors to point to a reality of which scientific theory has revealed only one aspect. It is the commitment to this reality that lends universal intent to a scientist's most original and solitary thought. By acknowledging this frankly we shall restore science to the great family of human aspirations by which men hope to fulfill themselves in the world community as thinking and sentient beings. For our problem is to discover a principle of differentiation and yet relationship lucid enough to justify and to purify scientific, philosophic and all other knowledge, both discursive and intuitive, by accepting their interdependence. This is the crisis in consciousness made articulate through the crisis in science. This is the new awakening.

Each volume presents the thought and belief of its author and points to the way in which religion, philosophy, art, science, economics, politics and history may constitute that form of human activity which takes the fullest and most precise account of variousness, possibility, complexity and difficulty. Thus *World Perspectives* endeavors to define that ecumenical power of the mind and heart which enables man through his mysterious greatness to re-create his life.

This Series is committed to a re-examination of all those sides of human endeavor which the specialist was taught to believe he could safely leave aside. It attempts to show the structural kinship between subject and object; the indwelling of the one in the other. It interprets present and past events impinging on human life in our growing World Age and envisages what man may yet attain when summoned by an unbending inner necessity to the quest of what is most exalted in him. Its purpose is to offer new vistas in terms of world and human development while refusing to betray the intimate correlation between universality and individuality, dynamics and form, freedom and destiny. Each author deals with the increasing realization that spirit and nature are not separate and apart; that intuition and reason must regain their importance as the means of perceiving and fusing inner being with outer reality.

World Perspectives endeavors to show that the conception of wholeness, unity, organism is a higher and more concrete conception than that of matter and energy. Thus an enlarged meaning of life, of biology, not as it is revealed in the test tube of the laboratory but as it is experienced within the organism of life itself, is attempted in this Series. For the principle of life consists in the tension which connects spirit with the realm of matter, symbiotically joined. The element of life is dominant in the very texture of nature, thus rendering life, biology, a transempirical science. The laws of life have their origin beyond their mere physical manifestations and compel us to consider their spiritual source. In fact, the widening of the conceptual framework has not only served to restore order within the respective branches of knowledge, but has also disclosed analogies in man's position regarding the analysis and synthesis of experience in apparently separated domains of knowledge, suggesting the possibility of an ever more embracing objective description of the meaning of life.

Knowledge, it is shown in these books, no longer consists in a manipulation of man and nature as opposite forces, nor in the reduction of data to mere statistical order, but is a means of liberating mankind from the destructive power of fear, pointing the way toward the goal of the rehabilitation of the human will and the rebirth of faith and confidence in the human person. The works published also endeavor to reveal that the cry for patterns, systems and authorities is growing less insistent as the desire grows stronger in both East and West for the recovery of a dignity, integrity and self-realization which are the inalienable rights of man who may now guide change by means of conscious purpose in the light of rational experience.

The volumes in this Series endeavor to demonstrate that only in a society in which awareness of the problems of science exists, can its discoveries start great waves of change in human culture, and in such a manner that these discoveries may deepen and not erode the sense of universal human community. The differences in the disciplines, their epistemological exclusiveness, the variety of historical experiences, the differences of traditions, of cultures, of languages, of the arts, should be protected and preserved. But the interrelationship and unity of the whole should at the same time be accepted.

The authors of *World Perspectives* are of course aware that the ultimate answers to the hopes and fears which pervade modern society rest on the moral fibre of man, and on the wisdom and responsibility of those who promote the course of its development. But moral decisions cannot dispense with an insight into the interplay of the objective elements which offer and limit the choices made. Therefore an understanding of what the issues are, though not a sufficient condition, is a necessary prerequisite for directing action toward constructive solutions.

Other vital questions explored relate to problems of international understanding as well as to problems dealing with prejudice and the resultant tensions and antagonisms. The growing perception and responsibility of our World Age point to the new reality that the individual person and the collective person supplement and integrate each other; that the thrall of totalitarianism of both left and right has been shaken in the universal desire to recapture the authority of truth and human totality. Mankind can finally place its trust not in a proletarian authoritarianism, not in a secularized humanism, both of which have betrayed the spiritual property right of history, but in a sacramental brotherhood and in the unity of knowledge. This new consciousness has created a widening of human horizons beyond every parochialism, and a revolution in human thought comparable to the basic assumption, among the ancient Greeks, of the sovereignty of reason; corresponding to the great effulgence of the moral conscience articulated by the Hebrew prophets; analogous to the fundamental assertions of Christianity; or to the beginning of the new scientific era, the era of the science of dynamics, the experimental foundations of which were laid by Galileo in the Renaissance.

An important effort of this Series is to re-examine the contradictory meanings and applications which are given today to such terms as democracy, freedom, justice, love, peace, brotherhood and God. The purpose of such inquiries is to clear the way for the foundation of a genuine *world* history not in terms of nation or race or culture but in terms of man in relation to God, to himself, his fellow man and the universe, that reach beyond immediate self-interest. For the meaning of the World Age consists in respecting man's hopes and dreams which lead to a deeper understanding of the basic values of all peoples.

World Perspectives is planned to gain insight into the meaning of man, who not only is determined by history but who also determines history. History is to be understood as concerned not only with the life of man on this planet but as including also such cosmic influences as interpenetrate our human world. This generation is discovering that history does not conform to the social optimism of modern civilization and that the organization of human communities and the establishment of freedom and peace are not only intellectual achievements but spiritual and moral achievements as well, demanding a cherishing of the wholeness of human personality, the "unmediated wholeness of feeling and thought," and constituting a never-ending challenge to man, emerging from the abyss of meaninglessness and suffering, to be renewed and replenished in the totality of his life.

Justice itself, which has been "in a state of pilgrimage and crucifixion" and now is being slowly liberated from the grip of social and political demonologies in the East as well as in the West, begins to question its own premises. The modern revolutionary movements which have challenged the sacred institutions of society by protecting injustice in the name of social justice are here examined and re-evaluated.

In the light of this, we have no choice but to admit that the *un*freedom against which freedom is measured must be retained with it, namely, that the aspect of truth out of which the night view appears to emerge, the darkness of our time, is as little abandonable as is man's subjective advance. Thus the two sources of man's consciousness are inseparable, not as dead but as living and complementary, an aspect of that "principle of complementarity" through which Niels Bohr has sought to unite the quantum and the wave, both of which constitute the very fabric of life's radiant energy.

There is in mankind today a counterforce to the sterility and danger of a quantitative, anonymous mass culture; a new, if sometimes imperceptible, spiritual sense of convergence toward human and world unity on the basis of the sacredness of each human person and respect for the plurality of cultures. There is a growing awareness that equality may not be evaluated in mere numerical terms but is proportionate and analogical in its reality. For when equality is equated with interchangeability, individuality is negated and the human person transmuted into a faceless mask.

We stand at the brink of an age of a world in which human life presses forward to actualize new forms. The false separation of man and nature, of time and space, of freedom and security, is acknowledged, and we are faced with a new vision of man in his organic unity and of history offering a richness and diversity of equality and majesty of scope hitherto unprecedented. In relating the accumulated wisdom of man's spirit to the new reality of the World Age, in articulating its thought and belief, *World Perspectives* seeks to encourage a renaissance of hope in society and of pride in man's decision as to what his destiny will be.

World Perspectives is committed to the recognition that all great changes are preceded by a vigorous intellectual re-evaluation and reorganization. Our authors are aware that the sin of *hubris* may be avoided by showing that the creative process itself is not a free activity if by free we mean arbitrary, or unrelated to cosmic law. For the creative process in the human mind, the developmental process in organic nature and the basic laws of the inorganic realm may be but varied expressions of a universal formative process. Thus *World Perspectives* hopes to show that although the present apocalyptic period is one of exceptional tensions, there is also at work an exceptional movement toward a compensating unity which refuses to violate the ultimate moral power at work in the universe, that very power upon which all human effort must at last depend. In this way we may come to understand that there exists an inherent independence of spiritual and mental growth which, though conditioned by circumstances, is never determined by circumstances. In this way the great plethora of human knowledge may be correlated with an insight into the nature of human nature by being attuned to the wide and deep range of human thought and human experience.

Incoherence is the result of the present distintegrative processes in education. Thus the need for *World Perspectives* expresses itself in the recognition that natural and man-made ecological systems require as much study as isolated particles and elementary reactions. For there is a basic correlation of elements in nature as in man which cannot be separated, which compose each other and alter each other mutually. Thus we hope to widen appropriately our conceptual framework of reference. For our epistemological problem consists in our finding the proper balance between our lack of an all-embracing principle relevant to our way of evaluating life and in our power to express ourselves in a logically consistent manner.

Our Judeo-Christian and Greco-Roman heritage, our Hellenic tradition, has compelled us to think in exclusive categories. But our *experience* challenges us to recognize a totality richer and far more complex than the average observer could have suspected—a totality which compels him to think in ways which the logic of dichotomies denies. We are summoned to revise fundamentally our ordinary ways of conceiving experience, and thus, by expanding our vision and by accepting those forms of thought which also include non-exclusive categories, the mind is then able to grasp what it was incapable of grasping or accepting before.

Nature operates out of necessity; there is no alternative in nature, no will, no freedom, no choice as there is for man. Man must have convictions and values to live for, and this also is recognized and accepted by those scientists who are at the same time philosophers. For they then realize that duty and devotion to our task, be it a task of acting or of understanding, will become weaker and rarer unless guidance is sought in a metaphysics that transcends our historical and scientific views or in a religion that transcends and yet pervades the work we are carrying on in the light of day.

For the nature of knowledge, whether scientific or ontological, consists in reconciling *meaning* and *being*. And *being* signifies nothing other than the actualization of potentiality, self-realization which keeps in tune with the transformation. This leads to experience in terms of the individual; and to organization and patterning in terms of the universe. Thus organism and world actualize themselves simultaneously.

And so we may conclude that organism is *being* enduring in time, in fact in eternal time, since it does not have its beginning with procreation, nor with birth, nor does it end with death. Energy and matter in whatever form they may manifest themselves are transtemporal and transspatial and are therefore metaphysical. Man as man is summoned to know what is right and what is wrong, for emptied of such knowledge he is unable to decide what is better or what is worse.

World Perspectives hopes to show that human society is different from animal societies, which, having reached a certain stage, are no longer progressive but are dominated by routine and repetition. Thus man has discovered his own nature, and with this self-knowledge he has left the state of nonage and entered manhood. For he is the only creature who is able to say not only "no" to life but "yes" and to make for himself a life that is human. In this decision lie his burden and his greatness. For the power of life or death lies not only in the tongue but in man's recently acquired ability to destroy or to create life itself, and therefore he is faced with unlimited and unprecedented choices for good and for evil that dominate our time. Our common concern is the very destiny of the human race. For man has now intervened in the process of evolution, a power not given to the pre-Socratics, nor to Aristotle, nor to the Prophets in the East or the West, nor to Copernicus, nor to Luther, Descartes, or Machiavelli. Judgments of value must henceforth direct technological change, for without such values man is divested of his humanity and of his need to collaborate with the very fabric of the universe in order to bestow meaning, purpose, and dignity upon his existence. No time must be lost since the wavelength of change is now shorter than the life-span of man.

In spite of the infinite obligation of men and in spite of their finite power, in spite of the intransigence of nationalisms, and in spite of the homelessness of moral passions rendered ineffectual by the technological outlook, beneath the apparent turmoil and upheaval of the present, and out of the transformations of this dynamic period with the unfolding of a world-consciousness, the purpose of *World Perspectives* is to help quicken the "unshaken heart of well-rounded truth" and interpret the significant elements of the World Age now taking shape out of the core of that undimmed continuity of the creative process which restores man to mankind while deepening and enhancing his communion with the universe.

RUTH NANDA ANSHEN

Translator's Note

The present volume is in the main a translation of the author's *Schritte über Grenzen* (Piper Verlag, Munich, First Edition, 1970). Certain essays have been omitted, however, since they are already available in English in other collections, e.g., *Philosophical Problems of Nuclear Science* (1952) and *The Physicist's Conception of Nature* (1958). The two final essays were not included in the original German edition, and the last is published here for the first time. I should like to thank two indefatigable typists, Mrs. Eusebia Shifflett and Mrs. Judy Catto, for their help in preparing the manuscript, and Dr. Heisenberg himself for checking and approving the text.

Preface

The present collection of essays and addresses, which have sprung, directly or indirectly, from the author's concern with atomic physics, repeatedly leads beyond the frontiers of this domain. The reason lies in the universal character of the science of the atom. Anyone who takes it seriously, with all its consequences in philosophy, technology and politics, has no other choice, when reflecting on these implications, than to trespass far beyond the boundaries of the field of physics proper. The directions in which this occurs in each particular speech or essay are so various that a systematic arrangement according to subject matter could hardly be achieved. In collecting them, therefore, it has been decided simply to arrange them in chronological order, except that those essays in which particular personalities are dealt with have been placed at the beginning. The temporal ordering to some extent provides an account of the development that has taken place in the author's thought, even though the choice of themes has often been determined more by the fortuity of circumstances. The same thoughts therefore recur in different contexts of inquiry; there has been no attempt to eliminate repetitions.

The most important topics can perhaps be indicated by way of the following questions: Where is technology taking us, now that, with the liberation of atomic energy, it has burst all its previous bounds? What content of truth do scientific assertions possess? Can there be agreement about the results of research, and can this agreement contribute to understanding among nations? Are there relationships between modern science and modern art? What can we learn from modern science to assist in the solution of ancient philosophical problems? No attempt has been made to give a systematic answer to these questions. It has been a matter, rather, of simply pursuing one's thoughts outward from the field of atomic physics to those points where, owing to the great importance of the problems attacked, it has been impossible to remain within the boundaries drawn by scientific specialism.

The individual speeches and articles of this collection have for the most part been published elsewhere. Their assembly in book form may, however, make evident a feature which is not sufficiently visible in each particular text, namely, that experience drawn from the development of atomic physics has led almost automatically to a way of thinking, united in its basic presuppositions and differing essentially at certain points from earlier scientific thought. This unitary character stems from the fact that the pure either/or, whose rigor often fails to match the reality, is replaced by a complementary way of looking at things. This makes it easier to see a problem from different sides and to avoid talking prematurely of unbridgeable opposites. It is a question, here, not of obliterating the clear outlines of earlier scientific thinking but of a more subtle approach to the possibilities initially concealed in that thought. The individual essays may be regarded as attempts to apply this way of thinking to fields outside the more narrowly scientific sphere.

WERNER HEISENBERG

ACROSS THE FRONTIERS

I

The Scientific Work of Albert Einstein

Albert Einstein was the most celebrated scientist of our age. Never before, probably, in the history of the sciences has a pioneer been known in his lifetime to so many people, and his lifework been intelligible to so few, as with Albert Einstein and his theory of relativity. Nevertheless, this fame is entirely justified. For somewhat like da Vinci or Beethoven in the arts, Einstein stood at a turning point in science, and his writings were the first to give expression to this change; it therefore seems as if he himself had brought about the revolution of which we have been witnesses during the first half of this century.

Einstein's name immediately became known among physicists when between 1905 and 1907 he came forward in four different areas of physics with trail-blazing new discoveries. The first was the atomistic interpretation of the theory of heat. It had already been known for some time that fine particles of dust, visible only under good microscopes, perform continuous irregular motions in fluids, a phenomenon which, under the name of "Brownian movement," had excited interest among physicists. Together with the Polish physicist M. von Smoluchowski, Einstein was able to account for this movement through the thermal motion of atoms and molecules, and to show how from the irregular movement of the dust particles something could be learned about the thermal motion, and thereby indirectly about

First published in *Universitas,* Vol. X, 1955, No. 9, pp. 897–902 (Wissenschaftliche Verlagsgesellschaft m.b.H. Stuttgart).

the size, of the molecules as well. This work contributed decisively to the establishment of confidence in the atomic hypothesis of the new physics.

In a second investigation, Einstein took up the inquiries of the Dutch physicist Hendrik Antoon Lorentz into the electrodynamics of moving bodies. In 1902, the American physicist Albert A. Michelson had first shown, by his celebrated interference experiment, that the motion of the earth in space—or, as was then said, relative to the ether—is not perceptible by optical tests. Then, in 1904, on the basis of a mathematical analysis of the situation created by Michelson's experiment, Lorentz had evolved certain transformation formulae, the so-called "Lorentz transformation," from which he concluded that moving bodies appear to contract in a definite manner in the direction of motion, and that moving clocks show an apparent time which elapses more slowly than real time. On these assumptions Lorentz could certainly interpret Michelson's result, but his formulae seemed on the whole to be unintelligible in terms of physics and therefore unsatisfying. Here Einstein took a hand, and with one stroke of wizardry resolved all the difficulties. He assumed that the bodies really do contract in the direction of motion, that the apparent time of Lorentz's formulae is actually the true time and that these formulae therefore convey a new understanding of space and time themselves. He thereby created the basis of the theory of relativity.

It will always remain difficult to make it understood why such seemingly trivial innovations should have brought with them consequences of quite extraordinary importance. It must be emphasized to start with that these changes presuppose a mental achievement of a highly uncommon kind. Until then it was among the self-evident presuppositions of science that space and time are two qualitatively distinct schemes of order, forms of intuition, under which the world is presented to us, and having nothing directly to do with one another. In any case there seemed to be only one time, the same time everywhere in the universe, for all organisms and for all dead matter alike. The whole of physics had been conducted since Newton's day upon these self-evident assumptions, and the great success of this branch of science had to be regarded as proof that they were correct—or largely correct, as we would now more cautiously say.

Einstein had the uncommon courage to cast all these assumptions into question, and he possessed the mental power to think out how, upon somewhat different assumptions, one may also arrive at a consistent ordering of the phenomena. In so doing, by 1906 he had derived, as one of the most important consequences, the inertia of energy, or, as it is commonly but less accurately put, the equivalence of mass and energy. It was this outcome of the theory of relativity that first brought it recognition and acclaim among physicists. It had been known for some time, from inquiries into the properties of the electron, that the electromagnetic energy stored in it also contributes to its inertia, i.e., to its mass. But the relation between inertia and energy seemed to depend in a complicated way upon assumptions about the shape of the electron. Now, at a blow, the theory of relativity furnished clear and simple relations. To be sure, it seemed at the time rather much to have to assume, say, that a clock becomes heavier on being wound up, and in fact the amounts of mass involved here are normally too small to measure. For a very small amount of mass corresponds, in relativity theory, to a very large amount of energy. But in the meantime the connection between mass and energy has come to exhibit a grossly measurable effect. The difference in mass between, say, the nucleus of a uranium atom and its two fragments after fission becomes only too clearly evident as the energy liberated in the explosion of an atom bomb. On this point, therefore, there can be no further doubt about the correctness of the theory of relativity.

The period between 1905 and 1907 covers two more important contributions from Einstein, in the field of quantum theory. After Max Planck, in 1900, had advanced the hypothesis that the light from atoms is emitted or absorbed not steadily but unsteadily in finite quanta, he kept on trying, despite all apparent contradictions, to reconcile this assumption with the wave theory of light accepted ever since Christian Huygens enunciated it in 1678. Einstein put an end to such attempts, in that again, as in the theory of relativity, he utilized the difficulties by making them the heart of the theory, and confronted the wave theory with the so-called hypothesis of light quanta. On this hypothesis, light consists not of waves but of fast-moving tiny corpuscles, which could be regarded as bundles of energy, or as Einstein said, light quanta. Einstein knew very well that on such a view one cannot initially interpret the diffraction and interference of light; but he recognized that the light quanta belonged nevertheless, in some fashion then still not understood, to the phenomenon of "light." Not until much later, on the formulation of the quantum theory in the mid-twenties, did we learn how to understand correctly the relationship between the wave theory and the hypothesis of light quanta.

Finally, Einstein made successful use of the basic assumptions of quantum theory in order to explain the specific heat of solid bodies in its dependence on temperature. Here, too, Einstein's paper represented the first decisive advance over the older approaches of the classical theory of heat.

With these four contributions, which bear witness to a truly extraordinary power and capacity for concentration, Einstein's fame was established as one of the leading scientists of the age. The theory of relativity proved ever increasingly to be the firm foundation, unshakable even by a great deal of criticism, for the whole of modern physics.

The portion of relativity theory so far alluded to is contrasted, in physics, as the so-called special theory of relativity with a more general version which Einstein himself produced in 1916. In this later general theory of relativity we are concerned with the attempt to combine gravitational phenomena with mass relationships in the four-dimensional space-time universe. In these inquiries Einstein started from the experimentally established equality between heavy and inertial masses. In an exceptionally bold combination of his physical ideas with Georg Friedrich Riemann's views on geometry, he interpreted gravitational fields as a deviation of the geometry of the four-dimensional space-time continuum from the geometry of Euclid. By this hypothesis he was able not only to account for ordinary celestial mechanics but also to explain certain subtleties, until then not understood, in the motions of the planets around the sun. This was yet another of the grandest achievements of a thinker inclined to abstraction, to recognize that even here the assumption of a non-Euclidean geometry can be rendered consistent with the observed phenomena, and that perhaps—though our observations do not yet extend so far—the universe is finite in size.

In the last ten years of his life, Einstein published articles chiefly of a philosophical or political nature, which, though not directly a part of his scientific work, nonetheless complete the picture of a savant creatively active on the broadest intellectual scale. In physics, Einstein's achievements were in the highest degree revolutionary, their consequences reaching out far beyond the science to which they initially belong. Yet, however paradoxical this may sound, in important aspects of his nature Einstein's was a conservative mind. Through his years of development he had become wedded to the nineteenth-century belief in progress, and his essays reflected the picture of a world which, though because of human irrationality it was exceedingly imperfect, could yet become better and better if men were ready to break with their former prejudices and put trust in their reason. In spite of unhappy experiences, Einstein was not prepared to part with this hopeful dream.

In the political sphere this attitude found expression in an almost naïve belief in the possibility of solving political problems solely through goodwill. The nationalist values of the day were alien to him, militarism he hated, and he gladly confessed himself a pacifist in the hope, characteristic of the period, that conflicts among men could be abated if a reform of society were to put power in the hands of new supranational institutions and the national states be compelled thereby to abandon the resort to violence in war. This picture of a peacefully progressing world is reflected in several of his essays, and it is thus a tragic aspect of his life that Einstein, to whom war was hateful, should have been moved by the infamies practiced under Nazism to write a letter to President Roosevelt in 1939 urging that the United States vigorously set about the making of atomic bombs, and that the first of these bombs should have killed many thousands of women and children who were just as guiltless as those for whom Einstein was anxious to intercede.

It would be absurd to treat this episode as an occasion for doubting the purity of Einstein's will. Anyone who knew him is aware that here was an earnest, kindly, disinterested man endeavoring only to do what was right. But the incident shows how that nineteenth-century world view, from which Einstein had freed himself only in physics, was no longer equal to answering the political questions confronting us in our own day.

That Einstein was deeply perturbed by this development is shown in many of the essays from the later years of his life. It is certainly no accident that this perturbation at times became linked in his mind with dissatisfaction about the development of physics since the end of the twenties. So let us review once more the consequences of Einstein's scientific work in regard to physics. First, Einstein replaced the space-time picture of classical physics by a new and more correct picture; he thereby showed that the foundations of the old physics were not so fixed and immutable as people had assumed them to be. Einstein further believed that in relating geometry to the field of matter he was creating a new and firmer foundation which permitted, no less than the earlier one, an objective description of nature, independent of man.

Here, however, Einstein overestimated what was possible in his day. For once the foundations of natural description had been set in motion, even the power of an Einstein was no longer sufficient to hold them immutable. Under the final interpretation of the quantum theory, to which Einstein himself had contributed so much, it became apparent, at the end of the twenties, that matter, space and time were themselves not such fixed realities, independent of man, as the nineteenth century had taught and as Einstein also wished to assume. He was now no longer willing to acknowledge such a shifting of the foundations. Unconsciously, at least, he no doubt felt that such an interpretation of quantum theory would also lend weight to those intellectual tendencies which had tended to be discredited as "ideological superstructure," in comparison with the hard reality of matter; hence this development dismayed him.

In his last years, however, Einstein's uneasiness had assuredly given way, instead, to the resigned wisdom of old age, which comes serenely to accept that the world eventually changes so much that it can no longer be explained by the images of our youth. In a scientific pioneer whose thought has contributed more than any other's to the changing of that world, this seems to us a particularly endearing feature of his character.

II

Planck's Discovery and the Philosophical Problems of Atomic Theory

If we are to speak, in what follows, of the philosophical consequences of Planck's discovery, it is first necessary to raise the question as to how a particular scientific discovery can have anything at all to do with general philosophical problems. This is plainly possible only when, by means of the discovery, questions of a very general nature are posed or answered; questions not so much directed toward a particular field of science but concerned, rather, with scientific method as such, or with the basic presuppositions of all natural science. The celebrated example in physics which shows this to be possible is the Newtonian mechanics, which at the beginning of the modern age posed anew the question as to what actually can be meant by the terms "understanding" or "explanation" of nature. The extraordinary influence of Isaac Newton's *Principia* on the thought of the following centuries was founded, not on the special axioms or results of this Newtonian mechanics—such as the well-known formula that "force = mass × acceleration"—but on the fact that for the first time natural phenomena could be described as they take place in time, and hence on the demon-

A lecture delivered at the session of the Association of German Physical Societies in honor of the centenary of the birth of Max Planck, on April 25, 1958, in the Congress Hall in West Berlin. First published in *Jahrbuch 1958 der Max-Planck-Gesellschaft zur Förderung der Wissenschaft e. V.*, and later in *Wandlungen in den Grundlagen der Naturwissenschaft* (S. Hirzel-Verlag) 1959, pp. 26–52.

stration that such a mathematical description of nature is basically possible.

If particular discoveries in natural science can acquire influence in this fashion upon the thought of whole centuries, such influence is not exercised in the fact that the discoveries have provided, say, a decision between various contending philosophical systems or that they have furnished the secure foundation for a new system of that kind. The connection between science and philosophy can never be so close. Nor should the observations that follow be misconstrued to mean that, from the standpoint of quantum theory or atomic theory, a position is being taken for or against one of the earlier or present-day systems of philosophy. The scientist's interest in philosophical ways of thinking is of a different sort. He is interested primarily in the putting of questions, only secondarily in the answers. The way questions are put seems valuable to him if they have been fruitful in the development of human thought. In most cases the answers can only be a product of their period; as our knowledge of the facts is extended in the course of time, these answers are bound to lose significance. It would especially, in every way, run counter to the spirit of science if an attempt were made to elevate some particular set of answers into a dogma. So we must try, on the contrary, to learn without prejudice as much as we can, both from the new facts and from the old and the new ways of putting the questions.

After these preliminaries, we have now to put the question as to the philosophical significance of Planck's discovery. What questions, then, of a general kind have been created by a conclusion reached in regard to the highly specialized problem of thermal radiation? What can Planck's formula

$$\varrho_\nu = \frac{8\pi\nu^2}{c^3} \frac{h\nu}{e^{\frac{h\nu}{kT}} - 1}$$

mean to philosophy? The fundamental character of the novelty introduced by Planck into modern science in 1900 can perhaps be best made clear by remarking that it revived discussion of the problem over which Plato and Democritus had contended some twenty-five centuries ago, and which represented the crucial point of disagreement between these two philosophers.

Here we must take a glance at the history of Greek atomism. The systematic thought of the Greek natural philosophers from Thales to Democritus had finally led to the problem of the smallest parts of matter. In place of the Parmenidean antithesis of being and non-being, with its termination in paradox, Democritus had postulated the antithesis between the full and the empty, i.e., between atoms and empty space. The existent, according to Democritus, is present an infinite number of times, in the form of a minute, unchangeable and indivisible constituent of matter. The diversity of what happens in the world is attributed to the varying arrangement and motion of atoms in the void. Just as tragedy and comedy can be written out in the same letters, so the most diverse happenings, in Democritus' view, can be actualized by means of the same atoms. But as to the nature of the atoms, and why they are just so and not otherwise, there is no further inquiry. The atoms are the ultimate given; they are indivisible and unchangeable, that which truly exists, from which everything is to be explained but which is itself in no need of further explanation.

Plato also took over significant elements of the atomic theory. To the four elements—earth, water, air and fire—for him, there correspond four kinds of smallest particle. In Plato's view, these elementary particles are basic mathematical structures of high symmetry. The smallest particles of the element earth are conceived as cubes, those of the element water as icosahedra, those of the element air as octahedra and finally those of the element fire as tetrahedra. But for Plato these elementary particles are not indivisible. They can be broken down into triangles and again be built up out of triangles. From two elementary particles of air and one of fire, for example, an elementary particle of water can be constructed. The triangles themselves are not matter but still simply mathematical forms. Thus for Plato the elementary particle is not the ultimate given, unchangeable and indivisible; a further explanation is needed, and the why and wherefore of these elementary particles is referred by Plato to mathematics. The elementary particles have the form Plato ascribes to them because mathematically it is the simplest and most beautiful. The ultimate root of appearances is therefore not matter but mathematical law, symmetry, mathematical form. Contention about the primacy of the form, the image, the Idea, on the one hand, over matter, the materially existent, on the other—or conversely, that of matter over the image—in short, the quarrel between idealism and materialism, has repeatedly agitated the thoughts of men throughout the history of philosophy. To the scientist, the difference between the two conceptions may often appear to be of no great importance. But Plato himself felt the conflict to be so profound that he is said to have expressed the wish that the books of Democritus might be burned.

But what does Planck's discovery have to do with this ancient question? For nineteenth-century chemistry, atoms were given as the smallest parts of the chemical elements. They were no longer themselves an object of investigation. The element of discontinuity or unsteadiness, which had manifested itself in the atomic structure of matter, at first had to be accepted without explanation. But Planck's discovery made it obvious that this same element of unsteadiness also appears elsewhere, namely in thermal radiation, where it certainly cannot be regarded simply as a consequence of the atomic structure of matter. In other words, Planck's discovery made it easy to suppose that this feature of unsteadiness in natural occurrences, which finds independent expression in the existence of atoms and in thermal radiation, would have to be understood as the consequence of a far more general law of nature. At this point, therefore, Plato's notion makes a renewed entry into science, to the effect that a mathematical law, a mathematical symmetry, ultimately underlies the atomic structure of matter. The existence of atoms or elementary particles as the expression of a mathematical structure was the new possibility that Planck opened up by his discovery, and here he is touching upon basic problems of philosophy.

To be sure, the road to a real understanding of this connection was still very far off. Another quarter of a century had still to pass before, on the basis of Niels Bohr's theory of atomic structure, a consistent mathematical formulation could be given for Planck's theory of quanta. And even in so doing, we were still far from any complete understanding of the structure of matter.

Nevertheless, with Planck's discovery a wholly new type of natural law was recognized to be possible, and with this we arrive at more specifically physical questions. The laws of nature that had previously been given a mathematical formulation, as in Newtonian mechanics or the theory of heat, had contained as so-called "constants" only the properties of the bodies to which they were to be applied. They included no constants having the character of a universal standard of measurement. The laws of Newtonian mechanics, for example, could be applied in principle to the motion of a falling stone, the moon's orbit around the earth or the collision of an atomic particle. The same thing seemed basically to happen everywhere. Planck's theory, however, contained what is known as "Planck's quantum of action," and it posited a specific standard of measurement in nature. It was made apparent that phenomena where the resulting effects are very large compared with Planck's constant take place in a basically different way from those where the effects become comparable with Planck's quantum of action. Since the events of our everyday experience are always concerned with effects that are very large in comparison with Planck's constant, it was demonstrated that phenomena in the atomic field might display features that altogether defy our immediate intuition. It might be a matter of processes that, though still experimentally observable in their effects and rationally analyzable by mathematical means, would no longer allow us to form any image of them. The unintuitable character of modern atomic physics rests, in the last resort, on the existence of Planck's quantum of action, on the presence in nature's laws of a yardstick of atomic dimensions.

Only a few years after Planck's discovery, natural laws containing a similar constant of measurement were formulated for a second time. This second constant itself, the velocity of light, had already long been known to physicists. But its basic role, as a standard of measurement in natural laws, was first made understood through Einstein's theory of relativity. Relationships exist between space and time, the seemingly wholly independent forms of intuition under which we apprehend phenomena, and in the mathematical formulation of these relations the velocity of light appears as the characteristic constant. Our everyday experience is almost always concerned with processes of motion that take place slowly in comparison with the velocity of light. Hence it is not surprising if our intuition boggles in the case of processes occurring at speeds approaching the velocity of light. This measure, furnished by nature, tells us not of particular things in nature but of the general structure of space and time. But this structure is no longer directly accessible to our intuition.

Following recognition of the basic significance of the two universal constants of nature, Planck's quantum of action and the velocity of light, it was natural to ask how many such independent constants of nature there can actually be. The answer is that there must be at least three such universal constants, but that probably all other constants of nature can be traced back to these three by way of mathematical relations that are still in part unknown. That three such independent natural units of measure must exist is something most simply made clear to the physicist or technician by considering how the customary physical or technical systems of measurement all contain three such units of measure: the centimeter for length, the second for time and the gram for mass. If we wish to replace these three conventionally established units by natural ones, Planck's quantum of action and the velocity of light must be supplemented by yet another constant. The atomic structure of matter makes it natural to choose as our third unit a length of an atomic order of magnitude, for instance a length of the order of the diameter of simple atomic nuclei. But a precise formulation of this unit of length can be given only if we are able to express mathematically the natural laws in which it figures as an essential quantity. Again we would expect our intuitive concepts to be applicable only to phenomena occurring in spaces that are large in comparison with this atomic unit of length, while in the region of smallest length—as this constant has also been called—phenomena would occur in a manner essentially different from that of our accustomed world.

But with this consideration we are hastening far ahead of the developments as they have actually taken place during the past decades. Planck's discovery had at first given only a glimpse into the possibility of tracing the atomic structure of matter back to mathematically formulated laws of nature, i.e., to mathematical forms. Though at that time it was scarcely possible to frame an idea of what sort of mathematical forms would eventually be involved, atomic physics had at least been set a goal. The gaze of the scientist was directed to the still distant summit of atomic theory, whence it would be possible to discern not only the existence of elementary particles and all the atomic products composed of them but also indirectly the physical interconnections of the world in general, as a consequence of simple mathematical structures.

At this point the hopes of the atomic physicists encountered the wishes of Albert Einstein, who in the 1920s was evolving the plan of advancing from general relativity theory to a unified field theory. Ever since the appearance of Einstein's theory of gravitation, the coexistence of various seemingly independent types of force field had been felt to be unsatisfactory. Physicists had long since been acquainted with such fields of force as the gravitational field and electromagnetic forces. To these had been added in the present century the matter waves, which can also be described as force fields of the chemical bond, and finally the many different wave fields that quantum theory assigns to the various elementary particles discovered in the preceding decades. Einstein hoped that it would be possible to regard all these force fields as evidence of the locally varying geometrical structure of space and time, and by means of the relation between geometry and matter to trace them to a common root.

In attempting to do this, Einstein attached fundamental significance to general relativity's proposed interpretation of the gravitational field by means of a positionally dependent geometry, whereas the quantum-theoretical regularities discovered by Planck were felt to be of secondary interest. Einstein was unable to accept as final the wholly unorthodox mathematical formulation of Planck's quantum theory, of which more must be said later, since it did not correspond to his philosophical conceptions of the task of the exact sciences. He felt it disturbing that natural laws should have to relate not to objective processes but to the possibility or probability of such processes. To the atomic physicists, on the other hand, it was Planck's quantum theory itself that appeared as the real key to an understanding of the interconnections. Hence they had to try to push forward to a unified field theory by way of the quantum theory of elementary particles. The antithesis between force and substance, which had played a certain role in nineteenth-century natural philosophy, had long since been resolved in quantum theory in the mathematically analyzed dualism between wave and corpuscle, or between force field and material elementary particle, so that the road onward to a unified field and matter theory seemed basically to be open.

But before following this road, so far as it has proved to be passable, and thereby tracing out the development of the last ten years, we must enter once more upon the epistemological situation which had arisen in the 1920s through Planck's discovery and its precise mathematical formulation. We have already made reference to a new type of natural law, in which given units of measure occur in nature. Perhaps it would be more correct to say that it is a matter of mathematically formulable basic structures in nature; for the concept of law is already almost too narrow to embrace these very general connections. Two such areas of connection were mentioned: quantum theory and relativity theory. These two theories have brought about radical changes in our picture of the world, because they have made it clear that the intuitive ideas by which we apprehend the things of our everyday experience are valid only in a restricted area of experience, and that hence they are by no means among the unshakable presuppositions of natural science.

In quantum theory we are particularly concerned with the question of an objective description of physical processes. In earlier physics, measurement served as the means of establishing objective states of affairs that were independent of the measurement. These objective states of affairs could be mathematically described and their causal connection thereby strictly laid down. In quantum theory, the measurement itself is still an objective state of affairs, just as in earlier physics; but the inference from the measurement to the objective course of the atomic occurrence to be measured becomes problematic, since the measurement intrudes into the occurrence and can no longer be separated entirely from the occurrence itself. An intuitive description of atomic processes, such as would have been sought in the physics of fifty years ago, therefore becomes impossible. We can no longer apprehend natural processes in the atomic field in the same way as processes on a larger scale. If we employ the accustomed concepts, their applicability becomes limited by the so-called "uncertainty relations." For the further course of the atomic process we can usually predict only the probability. It is no longer the objective events but rather the probabilities for the occurrence of certain events that can be stated in mathematical formulae. It is no longer the actual happening itself but rather the possibility of its happening—the *potentia,* to employ a concept from Aristotle's philosophy—that is subject to strict natural laws.

This aspect of quantum theory has often been expounded and I am not anxious to treat it here in excessive detail. Nor do

I wish to go into the history of this development, which is primarily linked with the names of Niels Bohr, Max Born, Pascual Jordan and Paul Dirac; or to deal, likewise, with the development of wave mechanics by Louis de Broglie and Erwin Schrödinger.

If we regard the step from classical physics to quantum theory as a final one, and therefore assume that even in the future exact science will include in its foundations the concept of probability or possibility—the notion of *potentia*—then a number of problems from the philosophy of earlier ages appear in a new light, and conversely, the understanding of quantum theory can be deepened by a study of these earlier approaches to the question. We have pointed out the relation to the concept of *potentia* in the philosophy of Aristotle. But there also emerge a host of relationships to modern philosophy in its various systems, relations that can only be sketched very briefly here. We can make no claim to a thorough and careful treatment.

In the philosophy of Descartes, the antithesis between *res cogitans* and *res extensa* played a decisive part, and the bifurcation of the world expressed in this pair of concepts has had a most powerful influence on the thought of succeeding centuries. In the physics of quantum theory, this antithesis looks rather different from what it did before. It appears less abrupt, for this physics has forced us to think in different areas of connection which stand to each other in a relation that Bohr has described by means of the notion of "complementarity." The areas in question may exclude one another but they also supplement each other, so that only through the interplay between them does the full unity become apparent. How this is possible without the slightest obscurity is shown in the mathematics of quantum theory. In comparison with classical physics, the quantum theory therefore manifestly departs from that rather too abrupt Cartesian division of the world.

Kant allotted a central place in his philosophy to so-called "synthetic judgments a priori" and the a priori forms of intuition. In the new interpretation of quantum theory, the basic concepts of classical physics are also recognized, indeed, as a priori elements; to that extent the theory includes a considerable portion of the Kantian philosophy. But at the same time, only a relative significance is accorded to the a priori, since, in contrast to Kant's view, even the a priori concepts no longer rank as immutable foundations of exact science.

The elements of positivism in relativity theory and the quantum theory have often been pointed out. Ernst Mach's arguments especially have undoubtedly often enriched the development of physics since Planck's discovery. But this influence, too, should not be overestimated. In particular, the quantum theory, in the interpretation generally accepted nowadays, by no means considers sense impressions to be the primary given, as positivism does. If anything is to be described as a primary given, then in quantum theory it is the reality that can be described by the concepts of classical physics.

Since quantum theory arose in conjunction with the atomic theory, it also stands, despite its epistemological structure, in close relation to those philosophies which give matter a central place in their system. But the development of recent years (of which more anon) in fact reproduces very clearly—if we are to make any comparisons at all with ancient philosophy—the transition from Democritus to Plato. For it is Planck's discovery which actually gives us the indication that the atomic structures of matter can be apprehended in natural laws as the expression of mathematical structures.

In addition, the epistemological analysis of quantum theory, especially in the form which Bohr has given to it, contains many features reminiscent of the methods of Hegelian philosophy.

Finally, a variety of inquiries have been undertaken into the relation of quantum theory to logic. I am thinking particularly of the investigations of Carl Friedrich von Weizsäcker. It is obviously possible to conjoin the quantum-theoretical interpretation of atomic processes with an extension of logic, which perhaps will acquire a very general significance in the exact science of the future. We have now taken what is admittedly only a very hasty glance at the manifold relations between quantum theory and a series of diverse philosophical approaches, into which we cannot enter here in detail.

In conclusion, we must mention at greater length a more physical problem, which already leads to the development of the quantum and atomic theories in the present century. Relativity theory and the quantum theory have revealed certain basic structures in nature that were previously unknown. In relativity theory we are concerned with the structure of space and time; in quantum theory with the consequences of the fact that every measurement in the atomic field requires an act of intervention.

The structure of space and time disclosed in the special theory of relativity can be briefly described as follows: We can group together under the word "past" all those events of which, at least in principle, we can experience something; and under the word "future" all those other events upon which, basically at least, we can still exert some effect. As we intuitively picture the matter, these two domains of events are separated only by an infinitely brief instant of time which we call "the present moment." But we now know from Einstein's theory that this region of the present is finite, and that the farther the place of the events is distant from our own, the longer it endures in time, because effects can never be propagated faster than with the velocity of light. Hence there is a sharp spatio-temporal boundary between those events of which we can have experience and those of which we can no longer have experience, and another boundary between those events on which we can still have an effect and those on which we can no longer have an effect.

But the existence of such a sharp boundary does not fit in well with the structure of the physical processes revealed by quantum theory. We know from the uncertainty relations that to determine a position requires an intervention of ever-increasing sharpness, the more accurately it is to be carried out. An infinitely sharp determination of position would actually presuppose an infinitely large intervention, and so cannot be realized at all. Hence it is no longer surprising that the sharp boundary insisted upon in relativity theory should lead to incongruities in attempting a quantum-theoretical formulation of physical processes. Here again it is impossible to enter into the details; but the literature of theoretical physics over the last twenty-five years is replete with discussions of these incongruities and apparent contradictions, which through the so-called "divergences," e.g., the infinite self-energy of the electron, have for long made it impossible to give a satisfactory description of processes among the elementary particles. Thus quantum theory and relativity theory obviously cannot be combined without difficulties.

After what has emerged in recent years, we have every reason to suppose that success in combining the two theories can be achieved only if we also take into the field of consideration the third basic structure, which is bound up with the existence of a universal length of the order of 10^{-13} cm.

Let us begin here by saying a little about which physical phenomena we are dealing with. Chemistry had originally allotted one kind of atom to each of the various chemical elements. Ernest Rutherford's experiments and Bohr's theory had shown that the so-called "chemists' atom" consists of a nucleus and a shell of electrons. The nuclear physics of the 1930s has taught us to regard the atomic nucleus as a structure of protons and neutrons. So three major kinds of elementary particle—protons, neutrons and electrons—were finally recognized as the ultimate building blocks of all matter. But later experiments showed that there are many other kinds of elementary particles, which are primarily distinguishable from the three already mentioned by the fact that they can exist for only a short time, since they very quickly undergo radioactive decay, i.e., are transformed into other particles. The mesons and hyperons were discovered, and today we are acquainted with some thirty different kinds of elementary particles, of which the majority possess only a very short lifetime.

These findings gave rise to two important questions. First, are these elementary particles, in particular the protons, neutrons and electrons, really the ultimate, indivisible building blocks of matter, or must they too be regarded as once more made up out of smaller particles? And if they are the smallest building blocks, why do they not allow of further division? Second, why are there only these experimentally detected elementary particles, and why do they have just the properties they are observed to have? What natural laws determine their masses and charges, the forces with which they act on one another?

To the first question, present-day physics gives the definite answer that the elementary particles do really represent the ultimate smallest units of matter; and it offers an initially rather surprising reason for this. How are we to establish whether the elementary particles cannot be still further divided? The only method of deciding this is surely to try to split them up further with the most powerful forces. Since there are naturally no knives or other implements with which the division could be attempted, the only remaining possibility is to make the elementary particles bounce into each other at high speed. One can in fact bring about collisions between elementary particles of the highest energy. The great accelerators which today are being constructed in the most diverse parts of the world, e.g., in Geneva as a European cooperative enterprise, in America and in Russia, serve this very purpose. The cosmic radiation occurring in nature also produces collisions of this kind. In the process, the elementary particles do indeed get broken up, being often smashed into many pieces, but—and this is the surprising thing—the pieces are no smaller or lighter than the elementary structures that were dashed apart. For according to relativity theory, the high kinetic energy of the colliding particles can be transformed into mass, and is actually so used in generating new elementary particles. In reality, therefore, there is no real splitting of the elementary particles, but rather an engendering of new particles of this kind, from the energy of motion of the incoming particles. Thus Einstein's equation

$E = mc^2$ creates the possibility that the elementary particles known today are in fact already the smallest structures that exist.

At the same time, we recognize here that the elementary particles are all, as it were, made from the same stuff, namely, if you like, from energy. Here one can find echoes of the philosophy of Heraclitus, for whom fire is the basic element of which all things are made. Fire is at the same time the driving force that keeps the universe in motion, and in order to arrive at our present view we may perhaps identify fire and energy. The elementary particles of modern physics can be transformed into each other exactly as in the philosophy of Plato. They do not themselves consist of matter, but they are the only possible forms of matter. Energy becomes matter by taking on the form of an elementary particle, by manifesting itself in this form. Here there is an echo of the relation between form and matter that plays such a central role in the philosophy of Aristotle. And with this, too, we have already arrived at the second question: why are there just these particular elementary particles and no others?

This question is identical with that concerning the natural law which determines the properties of the elementary particles; and this law must contain the third natural unit of measure, the so-called "smallest length." The problems at issue here are certainly by no means solved as yet, but it is already possible to advance for discussion a proposal for such a theory of elementary particles, which will have to be tested and developed by research in the years to come.

We must first review here the advances made in recent years. Some fifteen years ago, Dirac in England pointed to the possibility of resolving the mathematical "divergence difficulties" of quantum field theory, briefly alluded to above, by introducing into the mathematical presentation a new imaginary unit, the square root of -1, or, to put it in more accurate mathematical language, by assigning an indefinite metric to the Hilbert space of quantum field theory. Such an introduction actually implies a radical change in the structure of the theory, and soon afterward Wolfgang Pauli in Zürich was able to show that such a theory cannot initially be given a physical interpretation. For the quantities, which otherwise signify in quantum theory the probability for the occurrence of an event, can become negative in Dirac's formulation, and a negative probability is a physically meaningless idea.

But then in Göttingen, some five years ago, we took a further look at these thoughts of Dirac's, in the hope that the mathematical formalism might allow of development in the following manner. As has been frequently emphasized, the basic equation for matter must contain that unit of measure which was introduced as a length of the order of 10^{-13} cm. Might it not be possible to employ the indefinite metric in such a way that the negative probabilities make their appearance only when we ask about physical behavior in spatial dimensions of the order of 10^{-13} cm; but that for questions relating to very much larger regions of space and time, all the calculated probabilities again automatically become positive, so that the formulae allow of a physical interpretation? By this method the difficulties would be removed, for it is impossible even to ask about processes in the smallest regions of space; what this means is that processes in minimal space-time regions cannot be directly observed, and that conclusions from the observations to these processes can no longer be drawn with the concepts of ordinary physics. Hence these processes elude any kind of intuitive description. This possibility was thereupon studied in detail in a mathematically simplified model—I refer here to investigations in Göttingen by Mitter, Kortel and Ascoli—and it turned out that such a use of Dirac's suggestions is actually possible without contradiction. It also emerged from these inquiries that even such a simplified model already displays very important features of that unified field theory which must constitute the goal of the investigation. Thus the electromagnetic fields were derived, for example, as a consequence of the matter field, and the matter field manifested itself in elementary particles displaying properties similar to those actually observed.

A most important contribution to the problem of matter was then furnished by the discovery of two Chinese physicists in America, Tsung-Dao Lee and Chen Ning Yang, that electromagnetic fields are associated in a wholly unexpected way with a spiral direction inherent in elementary particles. For example, there are positively charged "π mesons," so called, which undergo radioactive decay. The μ mesons, electrons and neutrinos arising from this breakdown exhibit a particular polarization, corresponding in direction, as it were, to a right-hand spiral. There are no positively charged π mesons, in whose breakdown the reverse spiral direction can be distinguished. But there are indeed negatively charged π mesons of the same mass, and in their breakdown it is precisely the reverse spiral direction that determines the polarization. By mirroring, there thus arises from a particle its appropriate "antiparticle," so called, which carries just the opposite charge. This discovery had particularly interesting consequences for an understanding of the properties of an elementary particle, whose existence Pauli had predicted some time previously from an analysis of the β-decay of elements, namely the neutrino. In studying these consequences, Pauli had lighted in the previous year on a particular transformation property, a hitherto unnoticed mathematical symmetry in the neutrino wave equation. And now since, as was emphasized in describing the Platonic bodies, mathematical symmetries play a particularly important part in the theory of elementary particles, one could be prepared for the possibility that the symmetry mentioned above might have a significance extending beyond the special neutrino equation.

The experimental material on elementary particles that has been gathered together over the past twenty years provides information, though somewhat indirectly, on the symmetry properties in the basic equations of matter at that point where it yields us the so-called "selection rules and conservation principles." If we know from experience which particles can transform themselves radioactively into which others, we can reason backwards from this to the symmetry properties of the particles and to the laws underlying them. In attempting to transform the aforementioned mathematical model of a theory of matter, developed in Göttingen, in such a way as to take account of the observed rules of selection, we hit upon an equation of which Pauli showed that it also contains the symmetry properties he himself had discovered. Moreover, the Turkish physicist Gürsey pointed out that this symmetry of Pauli's obviously reproduces a characteristic property of the system of elementary particles which had been discovered twenty-five years earlier, and had achieved mathematical formulation in the concept of "isotope spin" or "isospin," which I shall not explain further here.

With this we could state an equation that—to put it cautiously—at least at first sight seems capable of depicting all the properties of elementary particles that are known to us—as if it might already be the true equation of matter. It runs:

$$\gamma_\nu \frac{\partial}{\partial_{x_\nu}} \psi \pm l^2 \gamma_\mu \gamma_5 \psi \left(\psi^+ \gamma_\mu \gamma_5 \psi\right) = 0$$

Here ψ (a field operator dependent on the space and time coordinates) signifies matter; the γ_μs are simple mathematical quantities introduced by Dirac from the theory of linear transformations and l is the natural unit of length already frequently referred to. The velocity of light and Planck's constant no longer show up in the equation, because we are using these two basic quantities as units of measure, and have therefore set them equal to 1. Quantity l can naturally also be employed in like fashion as a unit of measure, and then no longer figures in the equation.

It must be stressed at this point that the equation initially represents a proposal, and that only by a far from simple mathematical analysis of its consequences, in comparison with the experimental findings, will we be able, some years hence, to make a definitive judgment on how far this equation will take us.

For the moment it is perhaps more important to study the intellectual possibilities which, originating in Planck's discovery, have led, through the development we have outlined, to the advances of recent years. How will physics look if the hopes of physicists are fulfilled in this respect? In addition to the three natural units of measure, the preceding equation contains nothing except the mathematical requirements for symmetry. Through these requirements everything else seems to be answered. The equation should properly be regarded only as a particularly simple statement of the symmetry requirements, but these requirements must be seen as the real core of the theory. Just as with Plato, it therefore looks as if this seemingly so complicated world of elementary particles and force fields were based upon a simple and perspicuous mathematical structure. All those connections which we otherwise know as natural laws in the various departments of physics should permit of derivation from this one structure.

In this respect the modern view naturally has a degree of rigor that was altogether remote from Greek philosophy, and to avoid misunderstanding it is also necessary to stress the profound differences between our present-day natural science and that of antiquity. In the first place there is an essential difference of method, in that we undertake systematic experiments and accept theories only when they really delineate the experiments in every detail. And another very important difference appears in the part played in physics, since Galileo and Newton, by the concept of time.

The elementary particles in Plato's philosophy acquired their symmetry from the so-called "space group," i.e., the group of rotations in three-dimensional space. There, then, it was a question of a static, immediately intuitable symmetry. Modern physics, however, incorporates time into its view of nature from the outset. Since Newton, physics has been directed upon the dynamics of the phenomena. It starts from the view that in this constantly changing world the enduring element cannot be geometrical forms, but can only be laws. Yet the laws, too, are at bottom only more abstract mathematical forms, though they relate to both space and time. An understanding of matter therefore seems possible for us only if we reason from the experiments to mathematically comprehensible structures, which relate to space and time in the same way.

Just as in Plato, the eventual theory of matter will be characterized by a series of important symmetry requirements, which we can already state today. These symmetries can no longer be elucidated simply by diagrams and pictures, as was possible with Plato's bodies, but they can be made clear through equations. I should like to mention some of the most important of these equations, although such depictions can naturally be intelligible only to the mathematician.

A first crucial property of symmetry will be the so-called "inhomogeneous Lorentz group," which forms the basis of the special theory of relativity. A somewhat simplified version runs:

$$x' - x_o' = \frac{x - vt}{\sqrt{1 - \left(\frac{v}{c}\right)^2}}\,; \qquad t' - t_o' = \frac{t - \frac{v}{c^2}x}{\sqrt{1 - \left(\frac{v}{c}\right)^2}}$$

A second, equally important group is that of the transformations in Hilbert space which leave the exchange-relations invariant. This group is the basis of quantum theory. An equally simplified version goes like this:

$$\psi'_i> = S_{ik}\psi_k>$$

Furthermore, the so-called "isospin group," and the group associated with the conservation of baryon number, will play a part which, as we now surmise from the inquiries of Pauli and Gürsey, is represented by the Pauli transformations

$$\psi' = a\psi + b\gamma_5 C^{-1}\psi^+ \qquad (|a^2| + |b^2| = 1)$$

$$\psi' = e^{i\chi\gamma_5}\psi$$

Finally, there are also important mirror symmetries, e.g., the invariance of the theory under reversal of the time signs and under simultaneous spatial mirroring and charge reversal. All these symmetries are depicted in the equation above, though whether in correct form only the future will tell.

A theory which, from a simple basic equation for matter, correctly renders the masses and properties of the elementary particles is also at the same time a unified field theory. The experimentally known fact that all elementary particles can be transformed into one another is an indication that it could scarcely be possible to single out one particular group of such particles, say, and find a mathematical representation for this group alone. Because of this finding, and because of the fundamental importance of the symmetry properties, any attempted theory of elementary particles, such as that contained in the foregoing equation, acquires a peculiar character of closedness. One finds structures so linked and entangled with each other that it is really impossible to make further changes at any point without calling all the connections into question.

We are reminded here of the artistic ribbon decorations of an Arab mosque, in which so many symmetries are realized all at once that it would be impossible to alter a single leaf without crucially disturbing the connection of the whole. And just as these decorations express the spirit of the religion from which they have arisen, so also, in the symmetry properties of quantum field theory, we see reflected the spirit of the scientific epoch inaugurated by Planck's discovery.

But at this point we stand in the middle of a development whose results will be viewed only a number of years hence. In the half century whose individual stages I have tried to portray to you, Planck's discovery has led to a point where we think we clearly recognize in outline what the goal is, namely an understanding of the atomic structure of matter by way of simple mathematical properties of symmetry. Even if the recent developments of which I have spoken are approached with all that skepticism which is one of the supreme duties of the man of science, it can certainly be said that we have lighted here upon structures of a wholly uncommon simplicity, completeness and beauty—structures that seem to us especially important because they no longer have to do with a particular field of physics but with the world as a whole.

The centenary of Max Planck's birth falls in a period which, compared with earlier ages, presents an extremely chaotic appearance in many fields, notably those of politics, art and standards of value. It is therefore reassuring, particularly in remembering so harmonious a personality as Max Planck, that at least in the one field to which Planck devoted his lifework we find nothing chaotic; that here, on the contrary, simplicity and perspicuous clarity are still as much the rule as they were in the days of Plato, Kepler or Newton.

III

Wolfgang Pauli's Philosophical Outlook

Wolfgang Pauli's work in theoretical physics allows us only at a few places to recognize the philosophical background from which it has arisen. To his colleagues, Pauli appears preeminently the brilliant physicist, always inclined to the most incisive formulations, who by significant new ideas, by an analysis of existing findings clear down to the last detail and by unsparing criticism of every obscurity and inexactitude in proposed theories, has decisively influenced and enriched the physics of the present century. If we wanted to construct a basic philosophical attitude from these scientific utterances of Pauli's, at first we would be inclined to infer from them an extreme rationalism and a fundamentally skeptical point of view. In reality, however, behind this outward display of criticism and skepticism lay concealed a deep philosophical interest even in those dark areas of reality or the human soul which elude the grasp of reason. And while the power of fascination emanating from Pauli's analyses of physical problems was admittedly due in some measure to the detailed and penetrating clarity of his formulations, the rest was derived from a constant contact with the field of creative spiritual processes, for which no rational formulation as yet exists. Very early in his career, Pauli had followed the road of skepticism based on rationalism right to the end, to a skepticism about skepticism, and he then tried to trace

First published in *Die Naturwissenschaften,* 46th yr., 1959, No. 24, pp. 661–663 (Springer-Verlag, Berlin-Göttingen-Heidelberg).

out those elements of the cognitive process that precede a rational understanding in depth. There are two essays in particular from which the essentials of Pauli's philosophical attitude may be gathered: an article on "The Influence of Archetypal Ideas on Kepler's Construction of Scientific Theories"* and a lecture on "Science and Western Thought."** From these two sources and from his letters and other pronouncements, we shall try to obtain a picture of Pauli's philosophical point of view.

A first central topic of philosophical reflection for Pauli was the process of knowledge, itself, especially that of natural knowledge, which ultimately finds its rational expression in the establishment of mathematically formulated laws of nature. Pauli was not satisfied with the purely empiricist view whereby natural laws can be drawn solely from the data of experience. He allied himself, rather, with those who "emphasize the role of intuition and the direction of attention in framing the concepts and ideas necessary for the establishing of a system of natural laws (i.e., a scientific theory)—ideas which in general go far beyond mere experience." He therefore sought for a connecting link between sense perceptions on the one hand and concepts on the other: "All consistent thinkers have come to the conclusion that pure logic is fundamentally incapable of constructing such a linkage. The most satisfactory course, it seems, is to introduce at this point the postulate of an order of the cosmos distinct from the world of appearances, and not a matter of our choice. Whether we speak of natural objects participating in the Ideas or of the behavior of metaphysical, i.e., intrinsically real things, the relation between sense perception and Idea remains a consequence of the fact that both the soul and what is known in perception are subject to an order objectively conceived."

The bridge leading from the initially unordered data of

* In C. G. Jung and W. Pauli: *Naturerklärung und Psyche,* Studien aus dem C. G. Jung-Institut, Vol. IV, p. 109. Zürich (Rascher-Verlag) 1952.

** In *Europa—Erbe und Aufgabe,* Internat. Gelehrtenkongress Mainz 1955. Ed. M. Göhring, Wiesbaden (F. Steiner) 1956, p. 71.

experience to the Ideas is seen by Pauli in certain primeval images preexisting in the soul, the archetypes discussed by Kepler and also by modern psychology. These primeval images —here Pauli is largely in agreement with the views of Jung—should not be located in consciousness or related to specific rationally formulable ideas. It is a question, rather, of forms belonging to the unconscious region of the human soul, images of powerful emotional content, which are not thought but are beheld, as it were, pictorially. The delight one feels on becoming aware of a new piece of knowledge arises from the way such preexisting images fall into congruence with the behavior of external objects.

This view of natural knowledge is notoriously derived in its essentials from Plato, and it penetrated into Christian thought by way of neo-Platonism (Plotinus, Proclus). Pauli seeks to clarify it by pointing out that even Kepler's conversion to the Copernican theory, which marks the beginning of modern natural science, was decisively affected by certain primeval images or archetypes. He cites this passage from Kepler's *Mysterium Cosmographicum:* "The image of the triune God is in the sphere, namely of the Father in the center, of the Son in the outer surface and of the Holy Ghost in the uniformity of connection between point and intervening space or surroundings." The motion directed from the center to the outer surface is for Kepler the emblem of creation. This symbol, most intimately associated with the Holy Trinity, and described by Jung as a *mandala,* finds an imperfect realization, for Kepler, in the physical world: the sun in the center of the system of planets, surrounded by the heavenly bodies (which Kepler still thought to be animate). Pauli believes that to Kepler, the persuasiveness of the Copernican system is due primarily to its correspondence with the symbol described and only secondarily to the data of experience.

Pauli considers, moreover, that Kepler's symbol illustrates quite generally the attitude from which contemporary science has arisen. "From an inner center, the mind seems to move outward in a sort of extraversion into the physical world, in which all happenings are assumed to be automatic, so that the spirit serenely encompasses this physical world, as it were, with its Ideas." Thus the natural science of the modern era involves a Christian elaboration of the "lucid mysticism" of Plato, in which the unitary ground of spirit and matter is sought in the primeval images, and in which understanding has found its place in its various degrees and kinds, even to knowledge of the word of God. But Pauli adds a warning: "This mysticism is so lucid that it sees out beyond many obscurities, which we moderns dare not and cannot do."

He therefore contrasts the outlook of Kepler with that of a contemporary, the English physician Robert Fludd, with whom Kepler had entered into a violent polemic about the application of mathematics to experience refined by quantitative measurement. Fludd is here the exponent of an archaically magical description of nature, of the kind practiced by medieval alchemy and the secret societies that arose from it.

The elaboration of Plato's thought had led, in neo-Platonism and Christianity, to a position where matter was characterized as void of Ideas. Hence, since the intelligible was identical with the good, matter was identified with evil. But in the new science the world-soul was finally replaced by the abstract mathematical law of nature. Against this one-sidedly spiritualizing tendency the alchemistical philosophy, championed here by Fludd, represents a certain counterpoise. In the alchemistic view "there dwells in matter a spirit awaiting release. The alchemist in his laboratory is constantly involved in nature's course, in such wise that the real or supposed chemical reactions in the retort are mystically identified with the psychic processes in himself, and are called by the same names. The release of the substance by the man who transmutes it, which culminates in the production of the philosopher's stone, is seen by the alchemist, in light of the mystical correspondence of macrocosmos and microcosmos, as identical with the saving transformation of the man by the work, which succeeds only 'Deo concedente.'" The governing symbol for this magical view of nature is the quaternary number, the so-called "tetractys" of the Pythagoreans, which is put together out of two polarities. The division is correlated with the dark side of the world (matter, the Devil), and the magical view of nature also embraces this dark region.

Neither of these two lines of development, taking their rise from Plato and Christian philosophy on the one hand and from medieval alchemy on the other, could later escape disintegration into opposing systems of thought. Platonic thought, originally directed toward the unity of matter and spirit, leads eventually to a cleavage into the scientific and the religious views of the world, while the spiritual trend determined by gnosticism and alchemy produces scientific chemistry on the one hand and, on the other, a religious mysticism again divorced from material processes, as in Jakob Böhme.

In these mutually divergent and yet cognate lines of spiritual development, Pauli discerns complementary relationships which have determined Western thought from the outset and which today, now that the logical possibility of such relations has become fathomable to us through quantum mechanics, are more easily intelligible to us than they were to earlier ages. In scientific thinking, which is especially characteristic of the West, the soul turns outward and asks after the why of things. "Why is the one reflected in the many, what is the reflector and what the reflected, why did not the one remain alone?" Mysticism, conversely, which is equally at home in both East and West, endeavors to experience the unity of things, in that it seeks to penetrate beyond multiplicity, which it treats as an illusion. The scientific pursuit of knowledge led in the nineteenth century to the limiting concept of an objective material world, independent of all observation, while at the end point of the mystical experience there stands as a limiting situation the soul entirely divorced from all objects and united with the divine. Pauli sees Western thought as strung out, so to speak, between these two limiting ideas. "There will always be two attitudes dwelling in the soul of man, and the one will always carry the other already within it, as the seed of its opposite. Hence arises a sort of dialectical process, of which we know not whither it leads us. I believe that as Westerners we must entrust ourselves to this process, and acknowledge the two opposites to be complementary. In allowing the tension of the opposites to persist, we must also recognize that in every endeavor to know or solve we depend upon factors which are outside our control, and which religious language has always entitled 'grace.' "

When, in the spring of 1927, opinions on the interpretation of quantum mechanics were taking on rational shape and Bohr was forging the concept of complementarity, Pauli was one of the first physicists to decide unreservedly for the new possibility of interpretation. The characteristic feature of this interpretation—namely, that in every experiment, every incursion into nature, we have the choice of which aspect of nature we want to make visible, but that we simultaneously make a sacrifice, in that we must forgo other such aspects—this coupling of "choice and sacrifice," proved spontaneously congenial to Pauli's philosophical outlook. In the center of his philosophical thinking here there was always the wish for a unitary understanding of the world, a unity incorporating the tension of opposites, and he hailed the interpretation of quantum theory as a new way of thinking, in which the unity can perhaps be more easily expressed than before. In the alchemistic philosophy, he had been captivated by the attempt to speak of material and psychical processes in the same language. Pauli came to think that in the abstract territory traversed by modern atomic physics and modern psychology such a language could once more be attempted:

> For I suspect that the alchemistical attempt at a unitary psychophysical language miscarried only because it was related to a visible concrete reality. But in physics today we have an invisible reality (of atomic objects) in which the observer intervenes with a certain freedom (and is thereby confronted with the alternatives of "choice and sacrifice"); in the psychology of the unconscious we have processes which cannot

always be unambiguously ascribed to a particular subject. The attempt at a psychophysical monism seems to me now essentially more promising, given that the relevant unitary language (unknown as yet, and neutral in regard to the psychophysical antithesis) would relate to a deeper invisible reality. We should then have found a mode of expression for the unity of all being, transcending the causality of classical physics as a form of correspondence (Bohr); a unity of which the psychophysical interrelation, and the coincidence of a priori instinctive forms of ideation with external perceptions, are special cases. On such a view, traditional ontology and metaphysics become the sacrifice, but the choice falls on the unity of being.

Among the studies to which Pauli was prompted by the philosophical labors just referred to, it was those on the symbolism of the alchemists which left particularly lasting traces behind, as can be seen on occasion from utterances in his letters. In the theory of elementary particles, for instance, he is delighted with the various intertwined fourfold symmetries, which he at once relates to the tetractys of the Pythagoreans; or he writes: "Bisection and lessening of symmetry, that's the poodle's core. Dividing in two is a very old attribute of the devil (the word 'doubtful' is supposed to have originally meant 'twofold')."* The philosophical systems from the period after the Cartesian bifurcation were less congenial to him. The Kantian employment of the "a priori" concept he criticizes in very decided terms, since Kant had used this expression for rationally fixable forms of intuition or forms of thought. He expressly warns that "one should never declare theses laid down by rational formulation to be the only possible presuppositions of human reason." Pauli, on the contrary, brings the a priori elements of natural science into intimate connection with the primeval images, the archetypes of Jungian psychology, which

* [An approximation to the play in the original on *"Zweifel"* and *"Zweiteilung"* For "the poodle's core," see *Faust,* Pt. I, 1322, where Mephistopheles first enters, materializing from the shape of an enormous dog.—Tr.]

do not necessarily have to be regarded as innate but may be slowly variable and relative to a given cognitive situation. On this point, therefore, the views of Pauli and Jung diverge from those of Plato, who looked on the primary images as existing unchangeably and independent of the human soul. But in each case these archetypes are consequences or evidences of a general order of the cosmos, embracing matter and spirit alike.

In regard to this unitary order of the cosmos, which still cannot be rationally formulated, Pauli is also skeptical of the Darwinian opinion, extremely widespread in modern biology, whereby the evolution of species on earth is supposed to have come about solely according to the laws of physics and chemistry, through chance mutations and their subsequent effects. He feels this scheme to be too narrow, and considers the possibility of more general connections, which can neither be fitted into the general conceptual scheme of causal structures nor be properly described by the term "chance." Repeatedly we encounter in Pauli an endeavor to break out of the accustomed grooves of thought in order to come closer, by new paths, to an understanding of the unitary structure of the world.

It goes without saying that Pauli, in his wrestlings with the "One," was also continually obliged to come to terms with the concept of God; and when he writes in a letter of the "theologians, to whom I stand in the archetypal relation of a hostile brother," this remark is certainly also seriously intended. Little as he was in the position of simply living and thinking within the tradition of one of the old religions, so equally little was he prepared to go over to a naïve, rationalistically grounded atheism. No better account could well be given of Pauli's attitude to this most general of questions than that which he himself has offered in the concluding section of his lecture on science and Western thought:

> I believe, however, that to anyone for whom a narrow rationalism has lost its persuasiveness, and to whom the charm of a mystical attitude, experiencing the outer world in its

oppressive multiplicity as illusory, is also not powerful enough, nothing else remains but to expose oneself in one way or another to these intensified oppositions and their conflicts. Precisely by doing so, the inquirer can also more or less consciously tread an inner path to salvation. Slowly there then emerge internal images, fantasies or Ideas to compensate the outer situation, and which show an approach to the poles of the antitheses to be possible. Warned by the miscarriage of all premature endeavors after unity in the history of human thought, I shall not venture to make predictions about the future. But, contrary to the strict division of the activity of the human spirit into separate departments—a division prevailing since the nineteenth century—I consider the ambition of overcoming opposites, including also a synthesis embracing both rational understanding and the mystical experience of unity, to be the mythos, spoken or unspoken, of our present day and age.

IV

The Notion of a "Closed Theory" in Modern Science

The physical interpretation of modern quantum theory has brought up certain epistemological issues concerning the truth in scientific theories as such. To understand the point of view from which we nowadays judge the claim of such a theory to be true, it will be helpful to go over the historical development, and to trace from it how the goals of scientific endeavor have altered in the course of the centuries. So before going on to discuss questions of principle, let us begin with a brief historical survey.

1. Let us consider the beginnings of modern science in the sixteenth and seventeenth centuries. Johannes Kepler sought to discern the harmony of the spheres in the motions of the stars, in individual phenomena, that is, of particular importance and sublimity; he believed that in so doing he was on the point of discovering the divine plan of creation. The thought of a complete mathematical permeation of all individual processes on this earth was entirely remote from his mind.

Isaac Newton was not content to establish individual laws of particular mathematical beauty. He simply wanted to explain the processes of mechanics; he also recognized that in practice this was a task quite beyond accomplishment. But he thought

Published in 1948, at the instance of Wolfgang Pauli, in the journal *Dialectica,* Neuchâtel, Switzerland.

he could establish the basic concepts and laws whereby an explanation of this sort should at least be possible in the future. Newton conjoined the basic concepts by means of a group of axioms that could at once be translated into the language of mathematics, thus creating for the first time the possibility of reproducing an infinite wealth of phenomena in a mathematical formalism. By means of calculation, the individual complex process could be understood as a consequence of the basic laws and hence "explained." Even if the process itself was still quite unobserved, its outcome could nevertheless be "predicted" from the initial conditions and physical presuppositions.

The working out of this mechanics by subsequent generations was so successful that it inspired the idea that in principle it should be possible to trace back all occurrences in the world to mechanical processes taking place even among the smallest parts of matter. It was no longer thought possible to doubt the correctness of Newtonian mechanics. But since in this mechanics one could calculate the entire future of the system from the initial conditions, an exact knowledge of all the mechanical details of the world would also, it was argued, make it possible in principle to achieve a complete calculation of the future. This idea, most clearly stated by Pierre de Laplace, shows that by the beginning of the nineteenth century the type of mathematically formulated natural law created by Newton had already wrought an extensive transformation in scientific thinking.

For the nineteenth century, then, mechanics was simply coextensive with exact science. Its task and field of application seemed unlimited. Even Ludwig Boltzmann still took the view that a physical process has been understood only when it has been mechanically explained.

The first breach in this scheme of things was made by James Clerk Maxwell's theory of electromagnetic phenomena. This theory gave a mathematical account of certain processes without tracing them back to mechanics. It was only natural that as a consequence a heated controversy should have broken out about whether Maxwell's theory was intelligible without mechanics. Many tried to interpret his theory mechanically by postulating a hypothetical substance, the ether. This struggle really came to a head with Einstein's discovery, in 1905, of the so-called "special" theory of relativity. It was there concluded that in virtue of the assumptions about space and time implicitly contained in it, Maxwell's theory could not be traced back to mechanical processes that obey the Newtonian laws. The inference seemed unavoidable that either Newtonian mechanics or Maxwell's theory must be false.

For the next decade or two, some scientists and philosophers went on dourly defending the standpoint of Newtonian mechanics in the guise of the ether model; eventually the dispute, like many another quarrel about world views, was actually dragged into the political arena. But the majority of physicists, on the basis of the experimental findings, acknowledged special relativity and Maxwell's theory to be correct. The Newtonian mechanics continued, however, to serve as a good approximation to the correctness of relativistic mechanics for those processes in which all velocities are small compared with the velocity of light. In the limiting case of small velocities, the relativistic mechanics does in fact transform into the Newtonian.

But the very assumption that Newton's theory was, strictly speaking, "false" misled many scientists into unwittingly carrying over a fundamental hypothesis from the nineteenth century into the new physics. For although at this period the incipient quantum theory was already threatening from afar the internal consistency of classical physics, the working out of field theory, especially in the general theory of relativity, had such successes to record that many physicists saw it as the task of future science to describe the phenomena of the world by means of the concepts of field theory, and hence by means of a unitary conceptual scheme. They even attempted to give a mathematical interpretation of the atomistic features of nature, as singularities in the solutions of the field equations, and the de Broglie-Schrödinger wave mechanics seemed at first to fit in with this ideal of a general field physics. The basic concepts of relativistic field theory were certainly more abstract than those of Newtonian mechanics and harder to render intuitively, yet they were still wholly in accordance with our need for an objective and causal description of the processes involved, and were thus felt to be universal.

2. Even this illusion was destroyed by the quantum theory. For in it the formal mathematical apparatus simply cannot be directly patterned on an objective occurrence in space and time. What we establish mathematically is only to a small extent an "objective fact," and in large part a survey of possibilities. Thus the statement "Here is a hydrogen atom in its normal state" no longer contains a precise statement about the path of the electron but tells us instead that if we observe this path with a suitable instrument, the electron will be encountered at point x with a certain probability $p(x)$. The classical concept can be meaningfully applied only if one sees to it beforehand that insuperable limits are set to their application by the uncertainty relations.

The situation thus created in quantum mechanics differs very characteristically in two ways from the position in relativity theory: first, through the impossibility of any simple objectification of the state of affairs depicted in the mathematics, and the directly connected fact that it cannot be rendered intuitively; secondly—and this difference is perhaps still more important—through the consequent necessity of making a broader use of the concepts of classical physics. To describe the atom, we can and must employ such concepts as path of the electron, density of the matter wave at a particular point in space, heat of dissociation, color, etc. These are all concepts that belong to classical physics, insofar as they are supposed to represent objective processes in space and time. By means of them we describe the result of an observation. The various concepts often stand in a "complementary" relation to one another; but we cannot replace them by, say, other intuitive concepts whose usage would not be restricted by uncertainty relations or complementarity.

The result is that we no longer say, "Newtonian mechanics is false and must be replaced by quantum mechanics, which is correct." Instead, we now adopt the formula "Classical mechanics is a consistent self-enclosed scientific theory. It is a strictly 'correct' description of nature wherever its concepts can be applied." Thus even today we still concede truth, and even strict and general validity, to Newtonian mechanics, except that by adding "wherever its concepts can be applied," we indicate that we consider the field of application of the Newtonian theory to be restricted. The concept of a "closed-off scientific theory" first derives in this form from quantum mechanics. In contemporary physics we know of essentially four great disciplines, which we can consider as closed-off theories in this sense. In addition to Newtonian mechanics we have Maxwell's theory and the special theory of relativity, then the theory of heat and statistical mechanics and finally (nonrelativistic) quantum mechanics, together with atomic physics and chemistry. We must now say something in more detail about what properties belong to a "closed theory," and what content of truth such a theory can have.

3. The first criterion of a "closed theory" is its internal freedom from contradiction. It must be possible for the concepts initially stemming from experience to be so specified and fixed in their relations by definitions and axioms that mathematical symbols can be annexed to these concepts, among which symbols a consistent system of equations evolves. The most celebrated example of such an axiomatization of concepts is provided by the opening chapters of Newton's *Principia*. The wealth of possible phenomena in the relevant field of experience in nature is then reflected in the wealth of possible solutions to that system of equations.

At the same time the theory must, in a certain fashion, "depict" experiences; i.e., the concepts of the theory must, as already said, be directly anchored in experience, must "mean" something in the world of phenomena. The problems surrounding this particular requirement have perhaps not yet been sufficiently discussed. For so long as concepts stem directly from experience, like those of everyday life, say, they remain firmly linked to the phenomena and change along with them; they are compliant, as it were, toward nature. As soon as they are axiomatized, they become rigid and detach themselves from experience. To be sure, the system of concepts rendered precise by axioms is still very well adapted to a wide range of experiences; but we can never know in advance how far a concept established through definitions and relations will take us in our dealings with nature. Thus the axiomatization of concepts simultaneously sets a decisive limit to their field of application.

The limits of this field can never, indeed, be exactly known. Only the discovery that certain new groups of phenomena can no longer be ordered by means of the old concepts tells us that at this point we have reached the limit. In Newtonian mechanics, for example, perhaps the first inklings of the presence of a limit can be seen in the work of Michael Faraday, who sensed that the concept of a "field of force" is more appropriate to electromagnetic phenomena than are the concepts of mechanics. But the limit was not actually reached until nearly a century later, with the discovery of the special theory of relativity.

Even when the boundaries of the "closed theory" have been overstepped and new areas of experience are thus ordered by means of new concepts, the conceptual scheme of the closed theory still forms an indispensable part of the language in which we speak of nature. The closed theory is among the presuppositions of the wider inquiry; we can express the result of an experiment only in the concepts of earlier closed theories. Hence an attempt is sometimes made to reckon the concepts of older, closed-off theories among the a priori presuppositions of exact science, and thereby to lend them an absolute character to a still higher degree. In so doing, one side of the relationship is indeed correctly described. But here at least a difference of degree will have to be acknowledged. Such basic forms of human imagination or thought as space and time, say, or the causal law, which have been utilized and applied for thousands of years, must rank as a priori in a higher degree than the relatively complex concepts of closed-off theories dating from the last few centuries. If, as the biologist Konrad Lorenz has attempted to do, we consider the a priori forms of intuition to be "inborn schemata," it is clear that the established concepts of a closed-off theory dating from recent centuries cannot, or at least cannot yet, be a priori.

What, then, finally, is the truth content of a closed-off theory? The foregoing remarks may be briefly summarized as follows:

a. The closed-off theory holds for all time; wherever experience can be described by means of the concepts of this theory, even in the most distant future, the laws of this theory will always prove to be correct.

b. The closed-off theory contains no perfectly certain statement about the world of experiences. For how far one may be able to grasp phenomena by means of the concepts of this theory remains in the strict sense uncertain, and can be seen only by success.

c. In spite of this uncertainty, the closed theory remains a part of our scientific language, and therefore forms an integrating constituent of our current understanding of the world.

At the end of this discussion, let us return once more to the historical processes that, beginning with the changes in the picture of reality at the conclusion of the Middle Ages, have ultimately given rise to the whole of modern physics. This development appears to us as a succession of intellectual constructs, "closed theories," which take shape, as if from a crystal nucleus, out of individual queries raised about experience, and which eventually, once the complete crystal has developed, again detach themselves from experience as purely intellectual structures that nonetheless forever illuminate the world for us.

To that extent, amid all the differences, the history of the development of physics appears not unlike the history of other intellectual fields, for instance the history of an art; for even in these other fields the concern is ultimately with no other goal than that of illuminating the world, even if it be the world within us, by means of intellectual constructs.

V

Festival Oration for the 800th Anniversary Celebrations of the City of Munich (1958)

Today we are commemorating the eight-hundredth anniversary of the city of Munich. I really think it a pity that the congratulations at our festival assembly here should be uttered by a scientist. For when the name of Munich is heard, who thinks of anything so prosaic as science? Such a name brings other images to mind. The Ludwigstrasse, bathed in sunlight from the Siegestor to the Feldherrnhalle; the view from the Monopteros across the flowery meadows of the English Garden to the Frauenkirche; *The Marriage of Figaro* at the Residenztheater; the Dürers in the Pinakothek; the train crammed with skiers on its way to Schliersee and Bayrischzell; and finally, the beer tent at the Oktoberfest arena, crowned with the Bavarian lion. All that is Munich. But what has it to do with science?

Another thought, however, may well have been decisive here. Eight hundred years is no great age for a city; and Munich, especially, is in every way a young city. A lot of people think that it was really founded only about a hundred years ago, at least as the gay, artistic city we all love. On the birthday of so youthful a celebrant, it is fitting to think first of the present and future, and only secondarily of the past; so first let us speak of the picture the city presents to us today, and of our hopes and wishes for its future. But present and future are to be governed,

An address delivered in Munich on June 14, 1958. First published in the Munich press, 1958.

so they say, by science and technology. There may be some truth in that; and if Munich today is already a city of a million people, where the traffic roars through the streets like a mountain torrent, this is certainly a consequence of the technical industry of her people.

So let us begin by trying to forget for a moment the beloved image of the past and take a look at the city as it confronts us today. Anyone approaching it, say, from the north—we will suppose it a bright day of Föhn weather, which brings the mountains in close to the city—will find against this familiar background new shapes in the city's skyline, alongside the old landmarks: tall administrative buildings, emphasizing the character of Munich as a capital city; modern apartment blocks towering skyward; huge industrial developments, bearing witness to the prosperity and diligence of those who work in them. Even at first sight, a new Munich greets the eye of one who knew it in the old days; a city of industry and people, shaped by the spirit of our times. As we travel on the autobahn the last few miles across the broad plain separating the city from the hills about the Amper, there glides by on our left the aluminum dome of the Garching atomic reactor, built by the College of Technology and brought into operation this year. In its very outward form, this building tells us that here the ways of our time are being pursued, that here there is to be research and instruction in one of the latest fields of science. In the town, where people like to take a cheerful view of even the most serious matters, this building has long borne the appellation "atom egg." At the point where the autobahn runs into the northern suburbs, separated only by the Schwabing brook from the first coppices of the English Garden, another modern building is rising, which will also be devoted to atomic research and will provide accommodation this autumn for the Max Planck Institute for Physics and Astrophysics. Here there is to be experimental and theoretical work on the latest problems of nuclear physics and astrophysics.

Hence there will be attempts here to fathom the natural laws governing the properties and behavior of the smallest particles of matter. What is thus learned about atoms will be applied to an understanding of the cosmic processes taking place on the surfaces of stars and in interstellar space. From these purely scientific investigations, useful applications should also follow for technology, since we are on the track of thermonuclear reactions that may one day play a crucial part in the generating of energy. Once we have succeeded in building power plants on the basis of these discoveries, they will be far less dangerous than existing atomic power stations and will also use very much cheaper fuel. The latest branches of nuclear physics may thus also come to have great economic importance.

The main building of the new Institute is a modern steel-frame structure, similar to the many great administrative blocks in the inner city, whose broad glass facings allow light to stream into the building. When the heavy flow of traffic has caught us up from the site of the Institute and carried us into the heart of the city, we encounter these steel-frame buildings at many points, on the Oskar von Miller Ring, say, or on the Maxburg or in other places on the ring road girdling the inner city. These broad, bright, often perhaps rather hard but still friendly glass façades seem wrought by an optimistic spirit. They seem bent on telling us that we live in a bright, alert and open age, and that the one thing that can certainly never happen again is even the tiniest air raid.

In these buildings yet another feature of Munich also comes to the fore, which is not only typical of the present-day city but was repeatedly to be seen in the old Munich: the combination of the economically important or useful with the artistic. These buildings have been fashioned by architects who can also make things of beauty. We always rejoiced that the buildings of the old days were not only directly pleasing to us in their own proportions, like a work of art, but also seemed, so to speak, to have grown into their surroundings, as if they could never have stood anywhere else. This harmony of the old has been destroyed in many places by the war; but even the new buildings are at pains to fit in with their surroundings, to complete them, so that an overall cohesion, a rhythm in the street pattern, becomes visible. It has not been equally successful everywhere; but the artistic feeling expressed in the broad street patterns of the Bavarian kings, in the Ludwigstrasse and the Maximilianstrasse, is also still evident at many points in the Munich of today.

The new age, however, as a result of its science and technology, has brought one decisive change about which almost no one can be happy: peace has vanished from the streets. Through many of the major thoroughfares, the traffic roars day and night, so that nearby houses have become virtually unusable as dwelling places. In this respect our city is no better off than other great modern cities. On the contrary, the relatively high level of prosperity generated by the expansion of industry in Munich is resulting in so many cars as to detract from our former *Gemütlichkeit*. If we try here to peer a little into the future—and it is the future of our still so youthful city that we were supposed to be talking about—we may fancy that modern technology, which has created these truly formidable difficulties, may yet in the long run also have remedies available for them. The last war has only accelerated a development that would have occurred anyway. The residential areas proper are being shifted outward into the open spaces and woodlands around the city, whose true center continues to be used for business, for administration and in general for work. Once this process has advanced still further, it may come about naturally that the automobile will be used only for transporation from the suburbs to the ring road around the inner city, and that this traffic will be concentrated in a few great arteries at whose ends enormous garage buildings will receive the vehicles, with the inner city proper being absolutely barred to motor vehicles and accessible only to pedestrians proceeding from the ring road or from various subway stations. In some large cities a similar evolution has already taken place, and so perhaps the further development of technology will be able, not indeed to diminish the traffic in Munich, but at least to steer it into calmer and more orderly channels. Fortunately, the making of proposals for accomplishing this is a matter not for the scientist but for the traffic planner, who in Munich certainly does not confront an easy task.

On surveying the development in this area during the postwar years, one gets the impression that possibly growth occurred a little too rapidly, that the wholly exceptional economic expansion after the war often made it necessary to adopt emergency solutions that ought not to remain the final answers. Here we come to the question of how far the new aspects of the city we have just been speaking of conform only to their own laws and how far they can still be reckoned a natural continuation of the past, an extension of old traditions. Is the new Munich we have so far been discussing in any way still the same city as the old?

We can begin by declaring that the new Munich, with its expansion of trade and commerce, in many respects adheres quite automatically, for example through its geographical position, to old traditions that had already shaped the face of the city before the art-loving kings of Bavaria gave the place its particular character. The lumber trade has been domiciled for centuries in Munich as a natural consequence of its situation at the foot of the Alps. The brewing of beer has always been associated with the extensive cultivation of hops in the suitably situated part of Upper Bavaria toward the Danube. But craftwork also came early, especially goldsmithing, brass founding and the making of stained glass; by the beginning of the nineteenth century, book-printing and duplicating techniques were playing an important role. The first locomotive was built in 1841; and more than a century ago the scientific discoveries and technical advances of Joseph von Fraunhofer, and later of Karl August von Steinheil, were creating the basis for an optical industry. The lack of industrial raw materials has done much to ensure that economic development has proceeded differently, and given a different appearance to the city of Munich, from that to be seen, say, in the great industrial centers of the Ruhr. This shortage led on the one hand to a preponderant emphasis on trade in the early days, and more recently on refining and finishing work, while on the other hand it has produced from the outset a different economic atmosphere. The mere scramble for profit has always counted for less in Munich. Profit was a means of obtaining a comfortable contentment in life; and wealth, as one report has it, "walked the streets in respectable bourgeois attire." The economic importance of the refining trades was perhaps also the primary reason for the interest the city has taken in the development of natural science during the past century.

It was quite a long time before this interest could come to the surface. Fraunhofer's labors, for example, were pretty well neglected at first. But with the arrival in Munich of the so-called "Northern Lights," and especially with the summoning of Justus von Liebig by King Maximilian II, the ice was broken. The name and influence of Liebig were sufficient to establish in Munich an important center of chemical research, which for many years exerted a most powerful influence on science and technology. The synthesis of indigo discovered by Adolph von Bayer in 1883 became the starting point of the entire German dye industry. In succession to von Bayer, the Munich colleges can list a galaxy of eminent names—Richard Willstätter, Heinrich Wieland, Hans Fischer and many others—testifying to the intellectual vitality of this scientific center. By these means, then, did science make its entry into Munich; and it is characteristic of the period that the rise of science was at first viewed primarily in direct association with economic progress and improvements in the standard of living. As the nineteenth century ended, the belief in progress, which we now find somewhat charmingly romantic, took possession of Munich's soul; but, thanks to the other spiritual forces at work in our city, it also took on a special coloring that was typical of Munich itself.

This specifically Munich-like way of believing in progress found its strongest and most enduring expression in the Deutsches Museum, the lifework of one of our town's most vigorous personalities, Oskar von Miller. The Deutsches Museum owes its origin to a remarkable blend of endeavors of a highly diverse kind. At first it was no doubt intended simply to arouse interest in the progress of technology, and this task it has accomplished with the greatest success, especially among the rising generation. What youngster, in his day, has not delighted in playing with the various mechanical models—not always to the pleasure of the attendants? The Museum has then sought to make this progress intelligible to us, by exhibiting its historical development and encouraging the spectator to interest himself in the mechanisms, and even to try improving them. Finally, however, it enters into the quarrel, which our age is the first to be fully aware of, between man and the technology he has created. In the middle of the nineteenth century, when the growth of mechanical engineering began, there were few indeed —mostly artists, painters and poets—who could sense the first invasion of extremely dangerous, demonic forces into human life. Some looked with dread toward the coming era; others attempted to face even this development with optimism, and to see the brighter side of it. We may recall in this connection the poetic exchange between Gottfried Keller, whose Munich days are enduringly commemorated in *Der grüne Heinrich,* and the Swabian romantic Justinus Kerner, who had bitterly lamented, in one of his poems, the destruction of nature by technology. Keller answered him in the same form, with a poem about the steam engine:

To build the city, speed the plow,
I see it, glittering, snort and strain,
While lads on earth have leisure now
To prosper and to sing again.

A hundred years from hence, a flight
Of wine-ships out of Greece I see
Come planing through the morning light—
*Who would not then a pilot be?**

But the problem was not yet to be mastered with such optimism alone. The Deutsches Museum, possibly without knowing it, seeks to step in at this juncture, precisely by putting the human side of the story into the foreground. Technology as the spiritual adventure of man: art and technology make contact at this point, and hence such an answer, or attempt at an answer, has suited the temper of the city of art on the Isar. The task of combining art and technology, even more almost than art and science, has remained a live one in Munich throughout the whole of the last hundred years. Hence there have always been the liveliest exchanges of ideas between the College of Technology and the Academy of Art. Tasks in common have not been wanting, of course, especially in architecture; for example, in the artistic layout of the changes wrought by technology in the urban landscape. It can perhaps be said with some justice that the city of Munich has taken more trouble than any other industrial city in the world to look at technology on its human side, the side where it also makes contact with art. It is typical of the mood of this city that a series of lectures delivered some years ago in the College of Technology, on the arts in the age of technology, should have drawn several thousand young men into the packed auditorium every day for a week.

The scientific life at the University of Munich has also been marked from the outset by its relation to art and to the forces

* *Ich seh' sie keuchend glühn, und sprühen,*
stahlschimmernd bauen Land und Stadt,
indes das Menschenkind zu blühen
und singen wieder Musse hat.

Und wenn vielleicht in hundert Jahren
ein Luftschiff hoch mit Griechenwein
durchs Morgenrot käm' hergefahren,
wer möchte da nicht Fährmann sein?

that flow from the hearts of men. Academic life here has been different, in a way difficult to describe, from life in the other German universities. When the oldest Bavarian university was transplanted, in 1826, from Ingolstadt by way of Landshut to Munich, it seemed doubtful at first whether it would be able to thrive at all in a soil so initially alien to exact knowledge. And even when there could be no more doubt about its thriving, especially after the summoning of the "Northern Lights" by King Maximilian II, the University continued always to stand, after a fashion, in the shadow of art; for though students came to Munich to study, they were surely attracted primarily by the glitter and spaciousness of the city of art, by Richard Wagner and his romantic King Ludwig II, by the beauty of the landscape and the nearness of the mountains and only as a second best by scholarship and science. Yet the academic life of Munich has also received the most powerful stimuli from this proximity of art, from the vitality of the Munich spirit. We need only to recall here the names of Friedrich Wilhelm von Schelling, Friedrich von Thiersch and Wilhelm Riehl, and in later days of Heinrich Wölfflin, the art historian, and Karl Vossler, the Romance linguist. This relation is equally traceable in the historical sciences, sustained at the University by a series of eminent scholars, of whom Heinrich von Sybel, Heigel, Riezler and Hermann Oncken can alone be mentioned here. Even the scientists and physicians brought to Munich by Maximilian II, the former student of Göttingen, were soon able to develop a wide range of academic activities, and were much helped in doing so by the lively atmosphere of the city. We have already referred to Liebig, who made Munich a center of chemical research. The physicists Georg Simon Ohm and von Steinheil founded a scientific tradition that has been continued by such world-renowned men of science as Wilhelm Conrad Röntgen, Max Planck, Ludwig Boltzmann, Wilhelm Wien and Arnold Sommerfeld. In medicine, Munich had for decades occupied a prominent position. Two leading figures of more recent times have achieved reputations far outside the circles of their profession: Friedrich von Müller in internal medicine, and the brilliant surgeon Ernst Sauerbruch.

Here I must end a list of glittering names which could still be greatly extended. If other universities have won fame as repositories of solid learning or as starting points for new lines of development in research, our academic studies in Munich have been marked above all by a human immediacy and vitality, which have been able to flourish with astonishing success in the soil of a very conservative outlook, rooted in the Catholicism of the local population. The sensuous delight of the Baroque churches in Bavaria has had its secular counterpart in the cheerfulness, one might almost say the gaiety, of scientific work in the colleges, and both have been connected in some way with the light that streams on a sunny day across the peaks and pastures of southern Bavaria. These links with the countryside and the mountains have percolated even into life at the institutes and seminars, as when my teacher, Sommerfeld, would take some of his young physicists up to the Institute ski hut on the Sudelfeld, there to combine skiing with scientific discussion; or when once, at carnival time, the topmost room in the massive tower of the Physical Institute was declared to be a ski hut, so that the tower could be climbed only from the outside by the mountaineers. And since we have already got to talking of carnivals and ski runs, it has also always been a feature of life at the University of Munich that we do not shut ourselves off there from the doings of the larger community of the city but seek out social ties and companionship. For the younger members, skiing and carnivals may well come first in this regard; for the older ones, a diversity of clubs and societies sees to it that without any outward formalities, the bonds uniting the academic and artistic life of the city and those who have the fortunes of each in their charge are kept as close as possible.

At this point we should recall the one element in the life of our city that has always commanded a special degree of interest in those who visit it. The much talked-of district of Schwabing, which stretches from the Academy of Fine Arts and the two technical colleges at its southern extremity to around the industrial developments at Freimann in the north, is well known as the home of artists, writers and eccentrics. I don't know if that is quite so much the case nowadays. Yet still traceable everywhere are evidences of the artist's life, as we know it from Keller's *Der grüne Heinrich,* of the protests against bourgeois smugness and philistinism, of the poets Henrik Ibsen, Franz Wedekind, Stefan George and their circles, and of the hardihood of such innovators in painting as Vasili Kandinsky, Franz Marc and August Macke—even though this activity itself has only partly survived the destruction of the Second World War. The studio windows in the garrets of many an old tenement bear witness that painting still goes on in Schwabing, but in the new modern apartment blocks the studios have become rarer. Many little restaurants, old and new, try to continue in Schwabing the tradition of the old artists' taverns, and to capture in their interiors whatever of Bohemia may still remain here. But the attempt does not always succeed, and too often a dubious whiff of Broadway is wafted from New York into the Schwabing atmosphere.

In spite of everything, Schwabing, too, will have a key part to play in the Munich of the future, which was to be our primary topic today; for it embodies an element in the intellectual life of the city, without which Munich could simply not be Munich, the city so many people love. It embodies, as it has done for a century past, the spirit of tolerance. The old-established artisans and office workers in Schwabing, who took in painters or writers as lodgers, knew only too well what they wanted. They had both feet firmly planted in the soil of Munich, and did not have all that much use for what was new. But from a mixture of curiosity, admiration, contempt and a very considerable amount of natural friendliness, they conceded oddity its due. They did not take it all that seriously, and a hard word or two would often be dropped about it, again not altogether seriously intended, but they gave the new and the out-of-the-ordinary the space it needed. So not only artists but fanatics and crank reformers as well were able to live in Schwabing. They were scoffed at, but the scoffing was friendly and hence harmless. If people in Schwabing said of somebody, "He's cracked," that did not imply rejection. It was said in a friendly way, and to be cracked was to some extent accounted a recognized way of life in Schwabing. This spirit of tolerance survives in the Schwabing of today. It reacts upon the technical colleges and art schools, and forms an indispensable counter to the conservative outlook of the old Munich.

Just as the intellectual vivacity and receptiveness to novelty of the Schwabing artists had a fruitful and enriching effect on life in the colleges and in the city generally, so this mutual live-and-let-live in Schwabing set the tone for the easygoing character of the city as a whole, and thus created the preconditions for a harmonious interplay of all its forces. To be sure, Schwabing was something more than just lively and tolerant. Anyone who was there during the early 1920s will recall it as a place overflowing with youthful enthusiasm and *joie de vivre,* filled to the brim with music and poetry, sustained by the power of a number of extraordinary men who were able to bring young people under their spell there. But such years are festival periods, which cannot last. What can be hoped is that even in the future, Schwabing will remain disposed toward spiritual adventure, open to everything new, but without taking it too seriously, open to art and poetry, yet without undue solemnity. It may go on changing but still preserve its tolerance and its spirit of liberty.

The promenade we began with, through the modern Munich of today, has insensibly carried us into the past as well. Not into the whole eight centuries whose scenes passed before us at the pageant yesterday evening, but still into the period whose traces we continue to encounter at every step in the Munich of the present, that relatively recent past whose spirit still mingles with the spirit of our own day, to bring forth that very picture of the city which is known to us all.

What then, in fact, is the essential nature of this city? Assuredly the foundation of this multifarious essence continues to be the conservative, Catholic spirit of its native inhabitants, despite the many Germans from other parts of the former Reich who have found shelter and employment here. This rough, sturdy breed of old Bavaria, which for many centuries alone made up the city, continues even now to determine its basic character. And if new arrivals here are at times apt to fancy that the Bavarian unites in his person the amiability of the Prussian with the exactitude and punctiliousness of the Austrian, still, we cannot wish this race of men to be other than they are. A feeling for all the arts is active among them. It can be seen by the handicrafts practiced everywhere in Upper Bavaria, by the folk music and the organs in the churches; by the delight in beauty of spectacle, the showy brilliance, to which so many fine Baroque churches bear witness. The delight in beauty also includes pleasure in the theater, in games and festivals; the colorful trappings of a hunting party in a Bavarian mountain village can be viewed as a foretaste of the greater and richer glories of the region's capital. Ludwig II, who on winter nights would go sleighing by torchlight through the Upper Bavarian villages in a splendid equipage, was the peasants' king. His castles were fairy castles, and that was why the people, with their fondness for fairy tales, adored their king.

The roughness, on the other hand, sees to it that old things are not too much altered and that nothing false can creep in. The Bavarian insists on his ancient rights, but once they seem secure, he is soon ready to reach agreement. Differences of opinion, if they ever actually arise, are sooner settled over a jug of beer than with the knives out. In the life of the city, the delight in noisy conviviality and crude merrymaking has found its place in the Oktoberfest, which is held every year in September. Even in the Munich of tomorrow, this festival, like the Fasching carnival, will long remain an established feature of the calendar.

The tendency to cling to the good old ways is also doubtless an ultimately determining factor in the political life of our city, even if it does not always seem so; even if the gatherings, which are again naturally held, by choice, in the beer cellars of Munich, can become a source for outbreaks of political passion. Munich is a lively city—that is one of its chief and best qualities. But it is the same with cities as with people: their virtues are also at once their weaknesses and burdens. And a city of life and movement did once become a headquarters of "the Movement." But who wants, at a birthday celebration, to talk of the celebrant's weaknesses! The past can bid us remember that even a flourishing city may be very quickly destroyed by political unreason. But the dangers now come from faraway quarters of the globe, not from Munich. So let us not be too concerned about the future here. Even in the Munich of tomorrow, there will be strong forces seeking to preserve the good old ways wherever they can be preserved.

Now this foundation of conservativism is overlaid by the intellectual vivacity of which Schwabing is commonly said to be representative. This mental liveliness comes to a considerable extent from the north. The "Northern Lights" in fact had a most enduring influence on the mental life of Munich, and even up to recently it has often been the "incomers"—Germans from other regions or migrants from abroad—who have stimulated the city's cultural life. This is no less true of the arts than it is of science and learning. But the characteristic feature of Munich's spiritual climate is that the two levels have combined. The new intellectual interests that have filtered into Munich from the north or from remoter parts of Europe have found here a fruitful soil, and have been able to expand and flourish to an almost unexpected extent. And the Bavarians who already lived here found the novelties, which at first they had rejected as alien, so captivating and alluring that significant local talents soon also made their appearance, to follow the example of the "Northern Lights" and then to join them in adventuring forward into new regions of the spirit. Thus a Russian, Kandinsky, discovered in Munich new possibilities for painting, but it was the Munich-born Franz Marc who helped to develop and further the new trend.

This bonding of the two levels, of conservative Bavarianism with the Schwabing responsiveness to outside stimuli, which has borne such a rich harvest in Munich, has been possible only because the bearers of new tidings were happy to fit in with the Bavarian way of doing things. They gave thanks for the fundamentally tolerant and friendly attitude of the Bavarians, by joyfully seeking to participate, wholly and without reservation, in the local way of life. Thus out of mutual indulgence there soon grew up that mutual respect and friendship which has brought forth in consequence so much that is of value. Perhaps we should even consider what has been done here in Munich as an example of what ought to be happening in our day throughout much wider regions of Europe and the whole world. For a great deal of good comes of trying, despite differences which at first seem insuperable, to live together somehow with goodwill, and perhaps through just such a collaboration of highly dissimilar forces to bring something new into existence.

But let us return to Munich. When I was a young man growing up here, I became acquainted with the artist's life only among a limited circle. But as a student at the University I was able to take a modest part in academic life, and the thing I remember most clearly about those days—at the beginning of the twenties—is how very much the life of the University and of the professors, who were often from the north, was carried on in the city of Munich proper, in contact with those who really belonged there. These professors did not dwell in ivory towers but lived as Munich folk among their fellow citizens. My teacher Sommerfeld, for example, who came from East Prussia as Lovis Corinth the painter did, had long been in the habit, before the physics colloquia, of sitting in the Hofgarten along with other physicists, old or young, and there drinking coffee as so many other Münchners do. The talk was of the physical and mathematical problems that were then arousing interest, and the marble table the coffee was served on at times became covered with long mathematical formulae. I was told in those days that Sommerfeld once had to break off his table-top calculations with a complicated integral that could not be worked out further, since only a few minutes remained before the start of the colloquium at the University. A few days later, when Sommerfeld was again drinking coffee there with his students and happened to sit down at the same marble table, the calculation with the complicated integral was still there, but the solution had been scribbled in a few lines underneath. Another mathematician at the University of Munich—I think it must have been Herglotz—had drunk his coffee at the same table and whiled away the time by solving the integral.

That is how science will also be carried on at Munich in the future. When the technical college's atomic reactor at Garching was inaugurated a few months ago, the gathering was not confined to a few dignitaries from the state, the city and the colleges. On the contrary, it was a popular festival, in which the inhabitants of Garching also joined to the full, and where Bavarian national dishes were consumed to the strains of a band arrayed in *Lederhosen* and plumes of chamois hair. So it will also be in the future; even when the old-time commerce and trade have been replaced by the latest technology, by atomic power stations and fusion reactors, and even when—but this is too much to hope for—the Munich traffic problem has at last been solved by rocket transport.

It might be feared that in a cultural life where cheerful conviviality plays so large a part there would be rather too little room left for true intellectual toil, for that highest concentration of powers which is necessary, after all, for original creation in either art or science. The unrest of our day does not, in fact, stop short at the gates of Munich. Here, as in all other major cities, it threatens the calm and solitude of the study, where the real wrestling with problems is done. But anyone familiar with the city's intellectual life will know, and not only from the results, that this solitude really does exist, that here the true peace of toil is repeatedly found. But it is often not the solitude of the study but that of a copse on a moraine outcrop in the foothills of the Alps or a bathing place on one of the Oster lakes. A solitude in which one must, indeed, withdraw for a season from human company, but where full contact with nature and its secrets, with all the beauty of the Munich countryside, is nonetheless retained.

That is why science and art in Munich have always kept an element of romance about them, and so, no doubt, it will remain in the future. Even the most abstract painting in Munich has taken light and color from the lakes and pastures of the sunlit uplands beneath the Alps. When Arnold Sommerfeld, who was so well acquainted with the mathematical rigor of classical physics, came up against the new and still unexplained connections in quantum theory, he got so excited about the mysterious whole-number relations in the experiments on spectral lines that his lectures reminded one of Kepler's rhapsodies on the harmony of the spheres. Even when his critics accused him of number mysticism and fanaticism, when people said of him, "If it's integers you want, go to Sommerfeld," they could not upset his joy; for in Munich we never take things altogether seriously, and yet again we do, and in the end it was Sommerfeld who had very much the best of it over his critics. Similarly, the biology pursued here in Munich repeatedly contains this element of romance; whether it has to do with the sense of direction in bees or with the habits and psychology of higher animals, the conversations with ducks, geese and dogs that are carried on at the new Max Planck Institute for Behavioral Physiology on the Ess-See, between Starnberg Lake and the Ammersee.

From such collaboration, then, of old and new, of tradition and adventure, has arisen Munich, the city that not only lies in the heart of Europe but has also won a place in the heart of all Europeans. On the eight-hundredth birthday of this city, we do not need to be troubled about its future. The conservative and pious old town will continue to remain open to everything new and will go on reaping the fruits of that tolerance which has always been one of its chief virtues; and though the new image may also change repeatedly, though science and technology may transform the life of the place, yet in another way everything will also remain as it was. Of the Munich of tomorrow it will also be true, what was once said by Stefan George, perhaps just a trifle too pompously, of the Munich of his day:

Walls where the ghosts still wander abroad unfearing,
Ground by the double poison still unbleached:
City of youth and people! Home is reached
*Only, it seems, when Our Lady's towers are nearing.**

* *Mauern wo Geister noch zu wandern wagen,*
Boden vom Doppelgift noch nicht verseucht:
Du Stadt von Volk und Jugend! Heimat deucht
Uns erst wo Unsrer Frauen Türme ragen.

VI

Science and Technology in Contemporary Politics

In a lecture on language in politics, Carl J. Burckhardt has expressed the opinion that in an age when the great watchwords and ideals of the past seem to have been devalued by endless misuse and robbed of their ordering power, science and technology might, despite all the dangers involved, take on this role of bringing order into our thought. We shall seek here in a few words to pursue this idea.

First appearances are certainly against the hope that Burckhardt expresses. Although science and technology are at present exerting the strongest influence on the shaping of the world and are penetrating its remotest quarters, more often to bring destruction rather than order, the foreground is mainly occupied by material interest, which by its nature can operate in either direction. It may just as well lead to a showdown between hostile power groups, destructive of all order, as to the formation of well-ordered economic areas. Science can promise to alleviate material want, to cure diseases, to bring victory over enemies; its catchword "purposiveness" is persuasive. But even purposiveness can lead into chaos, if the purposes are not themselves understood as parts of a larger pattern, a higher order of things. "Purposiveness is the death of humanity." In saying this at times, it is pointed out that any isolated purpose

First published in *Dauer im Wandel.* Festschrift zum 70sten Geburtstag von Carl J. Burckhardt (Verlag Georg D. W. Callwey) 1960, pp. 194–197.

detached from its context can lead to developments inconsistent with what is truly human, namely the cautious tracing out of connections extending beyond the human sphere. The axiom fails to apply only when the purposes themselves form part of a larger context, which in earlier ages was spoken of as the divine order.

Although science here can initially operate in either direction, Burckhardt has himself already pointed to the educative aspect of the commerce with science and technology. Modern development has brought innumerable people in all parts of the world to devote themselves carefully and conscientiously to the solution of some technical or scientific problem presented to them. Whether it be the road builder, the maker of precision tools or the aircraft constructor, whether it is chemical processes to be investigated in the human organism or new plants to be cultivated for human use, the individual assigned to the task must always go soberly and carefully to work. He cannot let himself be dazzled by prejudices or illusions, he must renounce all the simplifications that are often so dangerous in political life, if he is really to be equal to the responsibility entrusted to him and to discharge it with success. This obligation of care and sobriety is already among the ordering forces of our age. But it would scarcely be sufficient if science were not also able to arouse a feeling for those larger interconnections in which the order of our world is expressed.

To the superficial observer, it may seem at first that science and technology are dissolving into an ever more incomprehensible welter of special disciplines, in which the individual, though he can still work successfully within them, is no longer able to survey their overall connection. But on closer examination we perceive behind this a movement in the opposite direction. Through the process of ever-increasing abstraction, which is taking place before our eyes in the exact sciences and is also gradually invading wider regions of intellectual inquiry, very broad connections, hitherto closed to human consciousness, are becoming visible not only within a given science but also between different sciences. The development of modern mathematics may serve as an example. The concept of number was originally formed by abstraction from things to be met with in sense experience; the geometrical figures arise by abstraction from the relations encountered, say, in land measurement. The reckoning with letters rather than numbers, the introduction of the imaginary unit, the study of functions, all betoken higher stages of abstraction. In accordance with the particular abstract structures under consideration, various branches of mathematics have been distinguished, such as arithmetic, algebra, theory of functions, topology and so on. But in our own day, mathematics is attaining to an essentially still higher level of abstraction by forming *superordinate* concepts, in terms of which the various topics of mathematics appear only as special cases of application, and in which, therefore, very general connections—logical structures—are reflected, which are operative in all the special disciplines of mathematics. From such concepts as set, group, lattice and operator, there emanates a binding force that allows mathematics to appear as a unity in a far higher sense than before.

Similar developments are also apparent in modern atomic physics. Earlier, chemistry and physics were separate sciences, concerned with quite different aspects of nature; and physics itself was broken up into a series of different disciplines—mechanics, optics, the theory of electricity, the theory of heat, etc.—which again had various kinds of processes and laws as their subject matter. Our age has grasped that all these phenomena are connected by way of law, but that in order to recognize the larger connections it is necessary to press forward into regions of nature that cannot be directly experienced through the senses. Once the physics of the atomic shells was understood, the unification of chemistry and physics was complete; with the experiments on elementary particles now being conducted with the most elaborate resources of technology, the connections between all kinds of forces in nature are emerging, and the formulation of the laws involved requires a degree of abstraction previously unknown in science.

In biology we are beginning to understand that the control of biological processes in the organism is often linked with particular properties of certain complex substances at the level of atomic physics. Here too we are obliged to leave the realm of immediately perceptible living processes, in order to recognize the connections at work. It seems, therefore, that developments in many different fields of science and technology are running in the same direction: away from the immediate sensory present into an at first uncanny emptiness and distance, whence the great connections of the world become discernible.

It must be emphasized at this point, however, that the renunciation of living contact with nature that accompanies our penetration into the new fields, and the associated trend toward abstraction in science, represent no deliberately chosen goal. On the contrary, they involve a very painful sacrifice, which finds its justification only in our knowledge of the broader connections. And these general orders have become truly visible in the science of our day. Hence there should be no comparison drawn here with modern art, although the drive to abstraction is very clearly in evidence there. The broad connections under discussion in contemporary science can indeed become known for the time being only to the narrow circle of those who work in this field.

Yet even from this circle it is possible for influences to spread into human thought generally. For example, a feeling will gradually grow up that life on earth represents a unity, that damage at one point can have effects everywhere else, that we are jointly responsible for the ordering of life upon this our earth. From the cosmic distances to which man can penetrate by the means of modern technology, we see perhaps more clearly than from earth itself the unitary laws whereby all life on our planet is ordered. That at this point the chance arises of penetrating into that initially uncanny emptiness and distance which technology and science have led us to, not only with the mind but also with the heart, is shown delightfully in the tale by the French aviator Antoine de Saint-Exupéry. His little prince, who looks after his small planet, cleans out the volcanoes and waters a rose, lives in that distance, but yet learns that *"On ne voit bien qu'avec le coeur, l'essentiel est invisible pour les yeux."*

If we raise the question whether science and technology in our life today are producing forces of order that may shape life on earth as did the great ideals of the past, we must surely think in the first place of those broad connections which have first become visible to us in these most recent developments. In regard to the great political dangers of our age, we may hope for a diffusion of the feeling that was succinctly expressed at an international congress by a Russian physicist: "We are traveling on a space ship that has already been circling the sun for untold ages, and is voyaging in company with that great star through infinite space. Whence and whither we know not; but we are traveling together on the same ship."

VII

Abstraction in Modern Science

When contemporary science is compared with that of earlier periods, it is often asserted that in the course of its development this science has become ever more abstract; indeed that at many points nowadays it has taken on a positively repellent air of abstractness, which is only partially compensated by the great practical success which science can point to in its application to technology. I have no wish to enter here into the question of value that is often raised at this juncture. We shall not, therefore, be asking whether the older kind of science was more satisfying, when by loving attention to the details of natural phenomena it gave life to relationships in nature, and so made them visible; or whether, on the contrary, the enormous enlargement of technical possibilities that stems from modern research has irrefutably demonstrated the superiority of our own conception of science. This question of value will thus be left wholly aside from the outset.

Instead, we shall try to examine the process of abstraction in the development of science itself. So far as this can be done within the scope of a brief historical survey, we shall be looking into what actually happens when science, in manifest obedience to an inner compulsion, ascends from one stage of abstraction to another lying above it; and what the cognitive values are for whose sake this laborious road of ascent is trodden at all. It will

Lecture delivered during the congress of the Pour le Mérite Order for Sciences and Arts, Bonn, 1960. First published in *Reden und Gedenkworte,* Vol. 4. Heidelberg (Verlag Lambert Schneider) 1962, pp. 141–164.

turn out in so doing that in the various disciplines of the scientific domain, very similar processes occur in each case—processes that become more intelligible precisely when we compare them. When the biologist traces metabolism and reproduction in living organisms down to chemical reactions, when the chemist replaces an intuitive description of the qualities of his materials by a more or less complex formula for their constitution, when the physicist ends by expressing natural laws in mathematical equations, a development is always occurring whose prototype is perhaps to be seen most clearly in the history of mathematics itself, and whose inevitability must be inquired into.

We may begin with this question: What is abstraction, and what part does it play in conceptual thought? And we may formulate an answer somewhat as follows: abstraction represents the possibility of considering an object or group of objects under *one* viewpoint while disregarding all other properties of the object. The essence of abstraction consists in singling out one feature, which, in contrast to all other properties, is considered to be particularly important in this connection. As can easily be seen, all concept formation depends on this process of abstraction, since concept formation presupposes the ability to recognize similarities. But since total sameness never occurs in practice among phenomena, the similarity arises only through the process of abstraction, through the singling out of one feature while neglecting all others. In order to be able, say, to form the concept "tree," it has to be realized that there are certain features common to birches and firs, which can be singled out and thus grasped by means of abstraction.

The discerning of common features can be a cognitive act of the greatest significance. It must, for example, have been recognized at a very early date in man's history that in comparing three cows, say, and three apples, there is a common feature, expressed by means of the word "three." Here already, the formation of the concept of number represents a decisive step out of the realm of the world immediately given to our senses and into a network of rationally apprehensible structures of thought. The statement that two nuts and two nuts together yield four nuts remains true even when we replace the word "nuts" by "loaves" or the designation of any other kind of objects. It was therefore possible to generalize it and to clothe it in the abstract form: two and two are four. That was a significant discovery. Very early also, in all probability, the peculiar ordering power of this concept of number was recognized and will have contributed to the feeling or interpreting of individual numbers as symbols. From the standpoint of present-day mathematics, however, the individual number is of less importance than the basic operation of counting. It is this operation that generates the interminable series of natural numbers and thereby implicitly gives rise to all the relationships that are studied, for example, in the theory of numbers. With counting, we have obviously taken a decisive step into abstraction and are able thereby to make our way into mathematics and mathematical natural science.

At this point we are already able to study a phenomenon we shall repeatedly encounter later on in the various stages of abstraction within mathematics or modern science, and which can almost be described as a sort of "ground phenomenon" in the development of abstract thinking in science—although Goethe would certainly not have employed his term *Urphänomen* in this context. We might call it "the unfolding of abstract structures." The concepts initially formed by abstraction from particular situations or experiential complexes acquire a life of their own. They prove to be far more abundant and fruitful than we can initially perceive them to be. In later development they display an independent ordering power—in promoting the creation of new forms and concepts, in providing insight into their connection and also in somehow demonstrating their own value when we seek to understand the world of phenomena.

Thus, from the notion of counting and the simple calculating operations associated with it, there later arose, partly in antiquity and partly in modern times, a complex arithmetic and theory of numbers, which is really engaged merely in discovering what was postulated from the outset in the concept of number itself. Number, moreover, and the resultant theory of numerical relations, created the possibility of comparing lengths by measurement. From there it became possible to develop a scientific geometry, whose concepts already go beyond those of number theory. On attempting, in this fashion, to found geometry on number theory, the Pythagoreans had already run into difficulty over the relation of incommensurable lengths, and had thereby been driven to enlarge their stock of numbers; they were *bound,* in a sense, to invent the concept of an irrational number. Moving on from there, the Greeks arrived at the concept of the continuum and at the famous paradoxes later studied by the philosopher Zeno. We shall not, however, be entering here into the difficulties involved in this development of mathematics; we merely wish to point out the wealth of forms implicitly contained in the concept of number and capable of being extracted from it.

This, then, can occur in the process of abstraction: the concept formed by way of abstraction takes on a life of its own; it allows for the generation of an unexpected wealth of forms or ordering structures, which can later prove valuable in some way, even in understanding the phenomena around us.

This basic phenomenon has notoriously excited debate as to what the subject matter of mathematics really consists of. It can scarcely be doubted that in mathematics we are dealing with genuine knowledge. But knowledge of what? Are we describing, in mathematics, an objectively real something, which therefore also has existence, in some sense, independently of man? Or is mathematics merely a capacity of human thought? Are the laws we deduce therein merely statements about the structure of this human thinking? I shall not really broach this difficult problem here but will merely point to a consideration confirming the objective character of mathematics.

It is not improbable that on other planets—on Mars, say, or at any rate in other solar systems—there is also something akin to life; and we must reckon with the possibility that on some of these other bodies there are also living beings, in whom the capacity for abstract thought has developed far enough for them to have framed the concept of number. If this is so, and if these beings have appended a scientific mathematics to their concept of number, they will have arrived at exactly the same propositions in number theory as we have done. Arithmetic and number theory cannot basically appear otherwise to them than they do to us; their results are bound to concur with our own. If mathematics is to be reckoned as a set of statements about human thinking, then at all events it must be about thinking as such, not merely about human thought. So far as there is any thinking at all, mathematics must appear out there in the same way. We may compare this conclusion with another conclusion, drawn from natural science: on other planets, or on bodies at remoter distances, exactly the same laws of nature assuredly hold good as they do with us. That is not just a theoretical opinion, for we can see in our telescopes that the same chemical elements exist there as they do with us, that they enter into the same chemical combinations and emit light of the same spectral composition. But at this moment we shall not inquire whether this scientific assertion, founded on experience, has anything to do with the other previous assertion about mathematics, and what it has to do with this.

Let us return for a moment to mathematics before taking a look at the development of the sciences. Mathematics, in the course of its history, has repeatedly formed new and more comprehensive concepts, and has thereby ascended to ever higher levels of abstraction. The realm of numbers was extended to include the irrational and the complex numbers. The concept of function gave access to the field of higher analysis, the differential and integral calculus. The concept of a group proved equally fruitful in algebra, geometry and the theory of functions, and suggested the idea that it should be possible, at a higher level of abstraction, to order and understand the whole of mathematics, with its many diverse disciplines, under a unitary point of view. The theory of sets was developed as an abstract foundation of this sort for the whole of mathematics. The difficulties of set theory eventually compelled the move from mathematics into mathematical logic, which was carried through in the twenties, particularly by David Hilbert and his collaborators in Göttingen. On each occasion, the step from one level to the next had to be taken, because the problems could not really be solved, or at all events really understood, within the narrow field in which they had initially been posed. Only the linkup with other problems in wider areas first opened the possibility for a new kind of understanding, and so occasioned the formation of further and more comprehensive concepts. Once it had been seen, for example, that the axiom of parallels in Euclidean geometry could not be proved, the non-Euclidean geometries were developed.

But a true understanding was not arrived at until a vastly more general question had been asked: Can it be proved within an axiomatic system that this system contains no contradictions? Only upon asking this question was the heart of the problem reached. At the end of this development there stands, in our own day, a mathematics whose foundations can be discussed only in concepts of an extraordinary abstractness, wherein the relation to things of any experiential kind appears to have been totally lost. The mathematician and philosopher Bertrand Russell has said, "Mathematics may be defined as the subject in which we never know what we are talking about, nor whether what we are saying is true." (To explain the second part of this statement, we know only that our propositions are formally correct, not whether there are objects in reality to which they could be related.) But the history of mathematics was meant only to serve as an example whereby we might be able to recognize the inevitability of the development toward abstraction and unification. It now has to be asked whether anything similar has taken place in natural science.

Here I should like to begin with the science whose subject matter should stand closest to life and to that extent be perhaps the least abstract, namely biology. Under its old division into zoology and botany, it was in large measure a description of the many forms in which life confronts us upon earth. The science compared these forms, with the aim of bringing order into the initially almost boundless wealth of life phenomena and of seeking for laws or regularities in the realm of the organic. Thus the question naturally arose, from what viewpoint the various organisms could be compared, and hence what the common features might be that could serve as a basis for comparison. Goethe's inquiries, for example, into the metamorphosis of plants were directed to just such a goal. At this point, therefore, the first step toward abstraction was bound to take place. There was no longer any primary concern with individual organisms, inquiry being directed instead to those biological functions, such as growth, metabolism, reproduction, respiration, and circulation, which are characteristic of life.

These functions yielded the viewpoint whereby comparisons could readily be made even among organisms of very different kinds. Like the abstract concepts of mathematics, they proved unexpectedly fruitful. They evolved, after a fashion, a power of their own for the ordering of very extensive regions of biology. Thus the study of processes of inheritance gave rise to the Darwinian theory of evolution, which promised for the first time to interpret the wealth of diverse forms among organic life upon earth under a broad unitary point of view. The investigations into respiration and metabolism led automatically in their turn to the question concerning chemical processes in the living organism; they gave occasion to compare these processes with chemical reactions in the test tube. The bridge from biology to chemistry was thereby thrown open, while at the same time the question was raised whether chemical processes in the organism and in inanimate matter occur according to the same natural laws. Thus the question about biological functions was transposed into the other question, how these biological functions are materially actualized in nature. So long as attention was directed to the biological functions themselves, the mode of consideration was still perfectly appropriate to the mental world of men like Goethe's friend the physician and philosopher Carus, who had pointed to the close connection between functional occurrences in the organism and unconscious mental processes.

But with the question as to the material actualization of function, the bounds of biology proper were burst asunder. For it then became obvious that one can have a real understanding of biological processes only when one has also made a scientific analysis and interpretation of the chemical and physical processes corresponding to them. At this next stage of abstraction there was therefore initial disregard of all biological sense connections, it being asked only what physicochemical processes actually occur in an organism as correlates to the biological processes. In pursuing this path, we have arrived today at a knowledge of very general connections, which seem in quite unitary fashion to govern all living processes on earth and which can be expressed most simply in the language of atomic physics. As a particular example, we may mention the hereditary factors whose propagation from one organism to another is governed by the well-known Mendelian laws. These hereditary factors are plainly given in a material sense by the arrangement of four characteristic molecular fragments on the twin fibers of a chain molecule called deoxyribonucleic acid (DNA), which plays a decisive part in the building up of cell nuclei. The extension of biology into chemistry and atomic physics therefore makes possible a unitary interpretation of certain basic phenomena in biology for the whole world of living things on earth. Whether the life that may exist on other planets makes use of the same structures in atomic physics and chemistry is still undecidable at present, but perhaps we shall also know the answer to this question in the not too distant future.

In chemistry there has been a development similar to that in biology. I shall pick out just one episode in the history of chemistry that is characteristic of the phenomenon of abstraction and unification, namely the development of the concept of valency. Chemistry is concerned with the qualities of substances and investigates the question of how substances with given qualities can be changed into others with different qualities; how substances can be combined, separated and transformed. When a start was made with analyzing combinations of substances in a quantitative fashion, and thus with asking how much of the various chemical elements was present in the compounds in question, whole-number relations were discovered.

Before this, the idea of atoms had already been employed as a useful picture by which to think of the combination of elements. The starting point here was the familiar analogy: if we mix white and red sand, say, we get sand whose reddish color becomes brighter or darker according to the relative mixture. This was how the chemical combination of two elements was thought of, except that atoms were substituted for grains of sand. Since the chemical compound is more different in its properties from the constituent elements than the mixture of sand is from the two kinds of sand, it was possible to extend the picture, and to suppose that the different atoms initially arrange themselves in groups, which as molecules then yield the smallest units of the combination. The whole-number relationships of the basic substances in the various combinations could then be interpreted in terms of the number of atoms in the molecule. Such an intuitive interpretation was confirmed by the experiments, and they also made it possible to assign to the individual atoms a number of so-called "valencies," which symbolized the possibilities of combination with other atoms.

In the process, however—and this is the point we are concerned with here—it remained at first wholly unclear whether the valency should be pictured as a directed force, a geometrical property of the atom or as something else. For a long time, indeed, it had to remain undecided whether the atoms themselves were real material entities or merely geometrical auxiliary notions making possible the depiction of chemical processes in mathematical form. By mathematical depiction we mean here that the symbols and their rules of combination—in this case, for example, the valencies and rules of valency—are "isomorphic" to the phenomena in the same sense as that in which it is said, in the mathematical language of group theory, that the linear transformations of a "vector" are isomorphic to the rotations in three-dimensional space. From a practical viewpoint, and without mathematical terminology, this means: Can we utilize the idea of valency to predict what chemical combinations will be possible between the elements in question?

But whether, in addition, the valency was also something real, in the same sense in which a force, say, or a geometrical form can be regarded as real, was a question that could be left without an answer for a long time, since the decision was of no great importance for chemistry. Throughout the complicated process of chemical reaction, attention had been directed primarily to the quantitative relations of mixture, to the neglect of everything else; that is, the process of abstraction had been employed, and a concept had been obtained which made it possible to achieve a unitary interpretation and partial understanding of the most diverse chemical reactions. Only much later, in modern atomic physics, have we learned what sort of reality lies behind the concept of valency. Even today we cannot rightly say whether valency is actually a force, an electronic orbit, an indentation in the charge density of the atom or even the mere possibility of something of the sort. But for present-day physics the uncertainty no longer relates in any way to the matter itself but only to its mode of verbal expression, whose imperfection we are basically unable to remove.

From the concept of valency, then, it is only a short step further to the abstract formulaic language of modern chemistry, which enables the chemist to understand the content and results of his work in all departments of his science.

Down the channels of the inquiry that seeks for unitary understanding and is thus led to abstract concepts, the streams of information collected by the observational and experimental biologist or chemist are thus eventually discharged automatically into the wide realm of atomic physics. It therefore looks as though, if only by virtue of its central position, atomic physics should be comprehensive enough to provide, for all phenomena in nature, a basic structure to which phenomena can be related and by which they can be reduced to order. But even for physics, which here appears as the common foundation for biology and chemistry, this is by no means self-evident, for there are very many different physical phenomena whose inner connection is initially beyond our ken. So now we must also go into the development of physics. Let us start by taking a look at its earliest beginnings.

At the threshold of natural science in antiquity there stood, of course, the discovery of the Pythagoreans that, as Aristotle reports it, "Things are numbers." If we give a modern interpretation to Aristotle's account of the Pythagorean theory, it surely means that things or phenomena can be ordered, and to that extent understood, by associating them with mathematical forms. Yet this association is not thought of as an arbitrary act of our cognitive faculty but as something objective. It is said, for example, that "Numbers are the substantial essence of things," or that "The whole heaven consists of harmony and number." This initially refers, no doubt, only to the order of the world as such. For ancient philosophy, the world is a cosmos, not a chaos. Nor does the understanding of the world so obtained appear, as yet, to be all that abstract; astronomical observations, for example, are interpreted by means of the concept of the circle. The stars move in circles. On account of its high symmetry the circle is an especially perfect figure; circular motion is self-evident as such. But to deal with the complex movement of the planets it was necessary even then to add numerous circular motions—cycles and epicycles—in order to represent the observations correctly. This expedient was completely adequate for the degree of accuracy attainable at that time. Eclipses of the sun and moon could be predicted very exactly by Ptolemaic astronomy.

At the beginning of modern times, this ancient view was

confronted in Newtonian physics with a question: Does not the moon's motion about the earth have something in common with that of a thrown or falling stone? The discovery that there is something common here, to which attention can be directed while disregarding all other and profounder differences, ranks among the most momentous occurrences in the history of science. The common factor was disclosed through the formation of the concept of "force," which effects a change in the "amount of motion" of a body; in this particular case, the force of gravity. Although this concept of force still stems from sensory experience—from the sensations felt, say, on lifting a heavy weight—it has nevertheless already become abstract in Newton's axiomatics, in that it is defined by a change in the momentum and without reference to these sensations. With a few concepts, such as mass, velocity, momentum and force, Newton builds up a closed system of axioms, which is now to extend to the treatment of all mechanical processes of motion, without regard for any other properties of the bodies in question. As we all know, this system of axioms, like the notion of number in the history of mathematics, has subsequently proved to be extraordinarily fruitful. For more than two centuries the mathematicians and physicists have drawn new and interesting consequences from the Newtonian formulation, which we learn at school in the simple form "mass × acceleration = force." The theory of planetary motions, inaugurated by Newton himself, has been developed and refined by later astronomy. Rotary motion has been studied and explained, the mechanics of fluids and elastic bodies has been evolved and the analogies between mechanics and optics have been mathematically worked out.

Two points of view must be particularly emphasized in this connection. In the first place, if we ask only about the pragmatic side of science, and hence compare, for example, the achievements of Newtonian mechanics in astronomical prediction with those of ancient astronomy, Newtonian physics, to start with at least, will scarcely have shown itself superior to ancient astronomy. By a superimposing of cycles and epicycles, it was possible in principle to depict the planetary motions as accurately as could be wished. The plausibility of Newtonian physics was not therefore primarily due to its practical applicability, but rested upon the collective viewing, the unitary explanation, of a great variety of phenomena; on the synthesizing power that emanated from the Newtonian formulation. In the second place, if new fields of mechanics, astronomy and physics were opened up in the centuries that followed, the major scientific achievements of a series of researchers were certainly needed for that purpose, but the results were already embedded, though they could not at first be recognized, in Newton's formulation, just as the concept of number implicitly contains already the whole of number theory. Even if rationally endowed beings on other planets were to take the Newtonian formulation as a starting point for their scientific deliberations, they could obtain only the same answers to the same questions. To that extent, even in the development of Newtonian physics, we are concerned with that "unfolding of abstract concepts" which was alluded to at the outset of this essay.

In the nineteenth century, however, it then turned out that the Newtonian formulation was not in fact copious enough to produce mathematical forms corresponding to all observable phenomena. Electrical phenomena, for example, which especially since the discoveries of Luigi Galvani, Alessandro Volta and Michael Faraday had been at the center of attention among physicists, did not fit properly into the conceptual system of mechanics. Faraday, therefore, rejecting the theory of elastic bodies, created the concept of a field of force, whose temporal changes were to be investigated and explained independently of the motions of bodies. From such beginnings there later developed the Maxwellian theory of electromagnetic phenomena, and from that the relativity theory of Einstein and finally the general field physics, which Einstein hoped could be extended into the foundation of the whole of physics. The details of this development will not be enumerated here. Of more importance for our present purposes is simply the fact that in consequence of such developments, physics was by no means a unified affair at the beginning of this century. In contrast to the material bodies whose movements were studied in mechanics stood the forces moving them which as fields of force now represented a peculiar reality with its own natural laws. The different force fields lay ranged alongside each other, without coordination. The electromagnetic forces and gravitation, which had already long been known, and the forces of chemical valency, were joined in more recent years by the forces in the atomic nucleus and the interactions regulating radioactive decay.

This juxtaposition of different intuitive pictures and distinct types of force created a problem that science could not evade, because we are persuaded that nature, in the last resort, is uniformly ordered, that all phenomena ultimately take place according to nature's unitary laws. So ultimately it had to be possible to discover the common underlying structure within the different branches of physics.

Again by means of abstraction and the forming of more comprehensive concepts, modern atomic physics has come close to this goal. The seemingly contradictory pictures yielded in the interpretation of experiments in atomic physics initially had the effect of placing the concept of "possibility," of merely "potential reality," at the heart of the theoretical interpretation. The conflict between the material particles of Newtonian physics and the force field of the Faraday-Maxwell physics was thereby resolved; both are possible manifestations of the same physical reality. The opposition between force and matter had lost its essential meaning. Moreover, the richly abstract concept of merely potential reality has itself proved to be extraordinarily fruitful; it was only through it that the atomistic interpretation of biological and chemical phenomena first became possible. But the desired linkage between the different types of force fields has come about in the last few years simply through new experiments. To every type of force field there corresponds, in the sense of this potential reality, a particular sort of elementary particle: to the electromagnetic field there corresponds the light quantum or photon; to the chemical forces there correspond in some degree the electrons; to the nuclear forces, the mesons, and so on. Experiments with the elementary particles demonstrated that when very swiftly moving particles of this kind collide, new particles are produced; indeed, it seems that if only enough energy is available in the collision for the forming of new particles, elementary particles of every desired kind can be created. Thus the various elementary particles are all, so to speak, made of the same stuff—we can call it simply energy or matter—and can be transformed into one another. The force fields, too, can be translated from one to another; their internal connection is discernible directly in experiment. For the physicist, there remains the further task of formulating the natural laws whereby the transformations among elementary particles take place. These laws have to represent or reproduce, in a precise and therefore necessarily abstract mathematical language, what can be observed in the experiment. With the increasing quantity of data, therefore, which is provided to us by an experimental physics operating with the most ample technical means, the solution of this problem should not be unduly difficult. In addition to the concept of a "potential reality" related to space and time, a particular role seems to be played here by the requirement that effects cannot be propagated faster than the velocity of light. For the mathematical formulation, we are eventually left with a group-theoretical structure, a set of symmetry requirements, that can be represented by fairly simple mathematical formulae; whether this structure is sufficient for the depiction of experience can again be established only through the process of "unfolding" already frequently referred to. But for present purposes the details are unimportant. The connection of the different regions of physics seems basically to have been accounted for by the experiments of the last ten years; we believe we can discern in outline the unitary physical structure of nature.

At this point, however, we must now also point out the limited character—grounded as it is in the nature of abstraction—of the understanding of nature that can be reached in this way. If we begin by neglecting many important aspects in favor of the one feature whereby we are able to order the phenomena, we are confining ourselves to the working out of a basic structure, a sort of skeleton, which only the addition of a great wealth of further details could turn into a genuine picture. The connection between phenomenon and basic structure is in general so involved that it can scarcely be followed through in detail. Only in physics has the relation between the concepts we use to describe the phenomena directly and those which appear in the formulation of the natural laws been largely explained. In chemistry this has been accomplished only to a very modest extent, and in biology we are just beginning in a few areas to understand how the concepts stemming from our immediate acquaintance with life, which in fact retain their value unconfined, can be fitted in with these basic structures. For all that, the insight acquired through abstraction furnishes a natural network of coordinates to which phenomena can be related and by means of which they can be ordered. The understanding of the world so obtained is related to the knowledge we first hoped for and toward which we continue to strive, as the plan of a landscape visible from a very high-flying aircraft is related to the picture we can get of it by walking about and living there.

Let us return to the question posed at the outset. The drive to abstraction in science is thus ultimately based upon the necessity to go on asking questions, upon the striving for a unitary understanding. Goethe once lamented this in connection with his self-created concept of the "ground phenomenon." In the *Farbenlehre* he writes: "But when even such a primordial phenomenon is arrived at, the evil still is that we refuse to recognize it as such, that we still aim at something beyond, although it would become us to confess that we are arrived at the limits of experimental knowledge." (*Theory of Colours,* § 177; Eastlake tr.) Goethe has felt clearly that we cannot escape the step into abstraction if we go on questioning. What he refers to in the words "something beyond" is simply the next highest stage of abstraction. Goethe wishes to avoid it; we should confess the limits of our view and not overstep them, because beyond this boundary seeing becomes impossible, and the domain of constructive thinking divorced from sense experience begins. This region remained always strange and sinister to Goethe, chiefly, no doubt, because its boundlessness appalled him. The limitless expanse to be seen here could prove attractive only to thinkers of a wholly different type. Nietzsche remarks: "For many the abstract is a source of weariness—for me on good days an intoxication and a feast." But men who consider about nature go on questioning, because they wish to grasp the world as a unity, to understand the unitary way it is built. For this purpose they fashion ever more comprehensive concepts, whose connection with immediate sense experience can be perceived only with difficulty—although the existence of such a connection is an indispensable precondition for the abstraction still to convey any understanding of the world at all.

Having been able to survey this process over so wide a range in the field of science today, we find it difficult, at the end of such an examination, to resist the temptation to cast a brief look at other areas of intellectual life, such as art and religion, and to ask whether similar processes have occurred or are still occurring there.

In the field of fine art, for example, we are struck by a certain resemblance between what happens to simple basic forms in the development of a style of art and what we have here called the unfolding of abstract structures. As in science, we have the impression that along with the basic forms—such as the square and semicircle in Romanesque architecture—the possibilities for elaboration and refinement are already largely determined for the richer forms of the later period, and that the development of the style is therefore more a matter of unfolding than of new creation. A very important common feature is also to be found in the fact that one cannot invent such basic forms but only discover them. The basic forms possess a genuine objectivity. In science, they have to depict reality; in art, to proclaim the content of life during the epoch in question. Under fortunate circumstances one may discover that there are forms which do this, but one cannot simply construct them.

It is harder to decide about the opinion sometimes advanced, that the abstractness of modern art has origins similar to the abstractness of modern science and is somehow related to it in content. If the comparison is justified at this point, it has this significance: only by renouncing the immediate link with sensory experience has modern art gained the opportunity to present and make visible wider embracing connections that earlier art was unable to express. Modern art can reflect the unity of the world better than its predecessors. Whether this view of the matter is correct I am unable to decide. The development of modern art is also often interpreted in a different fashion: the present-day dissolution of old orders, e.g., of religious ties, is reflected in art in the dissolution of traditional forms, of which only particular abstract elements remain behind. If this be the correct interpretation, there is no connection with the abstractness of modern science. For in exchange for the abstractness of science, insight into very wide-ranging connections has in fact been obtained.

Perhaps it is also permissible here to mention yet another comparison, from the field of history. That abstraction arises from continuing to ask questions, from the striving for unity, can be clearly recognized from one of the most momentous occurrences in the history of religion. The concept of God in the Jewish religion represents a higher stage of abstraction, when compared with the idea of many different nature gods, whose activity in the world can be experienced directly. Only at this higher stage is it possible to recognize the unity of the divine activity. If we may follow Martin Buber at this point, the struggle of the upholders of Judaism against Christ was a struggle to preserve the purity of this abstraction, to maintain the higher stage, once it had been gained. But against this, Christ had to insist on the requirement that the abstraction should not divorce itself from life, that men must submit directly to the working of deity in the world, even if there are no longer any intelligible images of God. That this is to point out the major difficulty of all abstraction is again only too familiar to us from the history of science. Every natural science would be worthless if its claims could not be tested by observation of nature; every art would be worthless if it was no longer able to move men, no longer able to illuminate for them the meaning of existence. But it would not be sensible to let our gaze wander so far afield at this point, where after all our concern was only to make intelligible the trend toward abstraction in modern science. So here we must simply be content to assert that the modern science of nature is arranging itself naturally into an immense scheme of connection, which arises from the fact that men continue to ask questions, and that such questioning is the form in which they come to grips with the world about them, in order to perceive its unitary connection and in order to live within it.

VIII

Current Tasks and Problems in the Promotion of Scientific Research in Germany

I should like to preface this report on problems of scientific research with two episodes that throw a somewhat disturbing light on the questions I am about to raise.

I was visited by leading members of the Japanese Research Council, who wanted to consult with me about appropriate measures for the promotion of research. After some two hours of cordial conversation, the head of this delegation, a leading Japanese scientist, took me aside and, after many apologies, asked me whether, between ourselves, he might put me one more question which was very important to him. On my agreeing, he said: "After the First World War, Germany was economically in an almost hopeless condition. But only a few years later, say from 1920 onward, and in spite of economic want, she was leading the whole world in scientific research. After the Second World War, the country made a much quicker economic recovery than after the first. As early as 1950 the economic position of Germany was much healthier than might have been expected. But even now, some eighteen years after the war, the country still

This speech, which must be understood in the context of its time, was delivered on November 5, 1963, before deputies of the Bundestag in the Interparliamentary Study Group, Bonn. First published in *Universitas,* Vol. 19, 1964, No. 10, pp. 1009–1022 (Wissenschaftliche Verlagsgesellschaft m.b.H., Stuttgart).

plays only a subordinate role in scientific research. What is the reason for that?"

The second episode I want to recount took place at the Institute for Plasma Physics in Munich-Garching. We need for our experimental work there a large number of electrical condensers, which must meet very exacting technical requirements. We knew that a company in England supplied condensers satisfying our conditions. But since a sizable order was involved—three thousand condensers for several million marks—we were urged to give the order to a German concern. After the technical specifications and test procedures had been precisely laid down and a reputable German firm had agreed to take on the contract, the order was placed. The first three hundred-odd condensers arrived in due course. Even on delivery, 4 percent of them were leaky and defective. One condenser was dismantled, and the cause of failure turned out to be faulty construction and assembly. To be brief, there followed an unhappy sequence of tests and letters and proposed modifications and so on. After some months the firm wrote to inform us that they now realized that they could not fulfill the technical conditions laid down in the contract and that it would therefore have to be canceled. Much time had been wasted. Disappointed, we got in touch with the English company. In the meantime, the fittings for the condensers had been delivered to Garching, and the English firm's condensers did not fit the installations there. But without hesitation, the English firm offered to try manufacturing them to our specifications and to build us condensers of the right size. Our chief engineer went over to England for talks. Just three days later, a test specimen arrived at the Riem airport; it was subjected to the same tests previously applied to the German condensers and functioned perfectly. Since then, all the English condensers have been installed at Garching, and are at present giving complete satisfaction.

I have started my report with these two episodes because I believe they should cause us to think. But I do not want to misuse them by drawing hasty conclusions. As a scientist, one has a duty to be skeptical and honest. Skepticism compels me to say that from two such casually selected episodes it is not possible to draw any well-founded conclusions about the state of German research. On the other hand, I can vouch for the fact that these incidents really took place as I have related them.

Let us begin with this question: What is the meaning of science in modern life? In earlier days, art and science were the cultural adornment of life, an adornment which could be afforded in good times but which had to be dispensed with in bad, since other cares and obligations claimed priority. Brilliant cultural achievements and material prosperity were the outward signs of a nation's good fortune. But today this is all utterly different. Whether we approve it or not, our whole life depends on science to an extent that could never previously have been imagined.

We shall therefore have to apply ourselves to this aspect of life if we want a correct answer to the question about the exigencies of scientific research. Such research is no longer the cultural adornment of life—though it can also be that—but is at any moment the seed corn from which economic prosperity, the right organization of state affairs, the national health and much else are due to grow. Perhaps, therefore, a partial answer to the Japanese scientist's question would run as follows: "We Germans have fallen behind because in recent years we have attended less than other nations to this aspect of modern life, the central position of science in our world."

But it is not only the importance of scientific research that has increased; the objects and nature of scientific work have also changed. To the old, traditional major areas of science, where we Germans have repeatedly made contributions of the greatest value, have been added new ones, which have commonly sprung up on the boundary between old areas. At the boundary, for example, between biology, chemistry and physics, molecular biology has developed, in which discoveries of the utmost importance have been made during the last decade. The application of scientific or mathematical methods to areas of the social sciences, such as economics or political science, has led to interesting new lines of thought, whose significance only later developments will decide. The technology of electronic calculators has brought into being the science of cybernetics, which has also proved extraordinarily fruitful in biology, in the study of the nervous system of simpler organisms. In these new fields, we Germans have found ourselves less easily at home than other nations who practice science.

The younger generation of scientists, after all, which must naturally make the advance into new territory, was in our case almost wholly wiped out by the war; in 1945, the scientific work at the technical colleges and research institutes had at first to be revived by the older men, and they often understandably took up again the areas of research in which they had done much before the catastrophe of 1933. But the coming generation of scientists, now on its way, is eager to work in the new and still unexplored fields. In the wider world, too, the style of research work has changed greatly. In view of the importance attached to research, governments have assigned it much larger resources. Research work is no longer carried on by individual inquirers but often by whole teams of young scientists; and in some places the whole arsenal of modern technology is called in, regardless of expense, in order to achieve certain research goals. Our technical colleges are not equipped for this type of research, and even at the more flexible Max Planck Institutes, the adaptation to such methods is not easy to bring about. Our administration, too, has the greatest difficulties in adapting itself to the needs of this new research, as will be noted again later on.

This new style of research also implies that we can no longer do everything everywhere, that not all technical colleges can engage in all lines of research. We have to set up concentration points, to gather the specialists together in suitable centers, to pursue what is important and to leave out what matters less. The forming of concentration points requires careful preparation in suitable administrative agencies, in which science itself must have a decisive influence. For this purpose, there can be virtually no other method but that of careful consultation between the experts and the administrators. In recent years the various commissions in the Ministry of Science, the German research community and the Science Council have done outstanding work; I see no reason for making many changes here in the existing forms of organization. If in the future we are confronted with still higher demands, for example through international arrangements, perhaps these commissions will need to discriminate more closely than before among the expensive projects. We shall give special regard to those areas and topics of research in which the prospects of success are brightest or which for other reasons must rank as particularly important, while less important ones are cut back.

I could not, on the whole, regard it as a misfortune if a rather keen wind were to blow here. Selectivity often brings with it an improvement in quality as well. In particular, we shall have to think about the following: If new fields of science are to be attacked or new methods employed, then since unlimited financial resources and unlimited manpower can never be available, old enterprises must of necessity be abandoned. This requirement falls more heavily nowadays upon us Germans than on other countries, for our self-consciousness cannot be founded on the period after 1933. Hence it must be based on what was done before that time, and that of course consists of the old sciences and the old forms of administration. A second partial answer to the question posed at the outset would therefore run: We Germans are at the moment still too little disposed to abandon the old in order to facilitate the new. We must learn at this point to run greater risks, as it were. Scientific audacity has never been wanting in Germany in the past; we must learn it again, must seize upon the new, taking firm decisions about what seems to us important and what does not.

Now, speaking of the administrative aspect, I come to a part of my report that is particularly vulnerable to criticism: What can Parliament and Government do to improve matters? There are two highly controversial issues here: the level of public provision for research and the modernization of administration.

First a few figures, which I believe to be fairly reliable. In other industrial countries, such as America, Britain and France, the proportion of the national budget expended on nonmilitary scientific research is running at about 4 percent. In the Federal Republic it is some 1.7 to 2 percent, or approximately half of this. Such figures, as we all know, are problematic, because it is hard to total and then compare exactly similar amounts in the different countries. For example, it might be said of the Federal Republic that we ought not to start from the Federal budget but from the total of Federal and Länder budgets together. But not much is altered if we do. If we pay attention only to expenditure on scientific research, and thus reckon, as in other countries, that a major part of expenditure on higher education goes to instruction, scientific training, support of students, etc., and that only a small part of such expenditure is assignable to research in the proper sense of the word, we arrive once more at approximately the same figure of 1.7 to 2 percent for the Federal Republic. I believe, therefore, that we can with a good conscience take these figures as our starting point.

If one discusses these figures with well-informed members of the Bundestag or Federal administration, one often hears this objection: "Such figures have no really convincing effect on us. We would like to have particular well-founded scientific projects put before us, but not these general considerations, which are difficult to check. Where concentration points are in question and selections have to be made, that is precisely where we want to have information and the chance to participate in decision." This request is basically quite justified, and particular projects are in fact put forward, for example in the budget proposals of the Ministry of Science, and are carefully scrutinized by various committees. But in making such requests one should not set out from illusions. In establishing and disbursing the Federal budget it is generally necessary to proceed in the same way as an institute director, such as myself, has to do when together with my colleagues I plan the budget of the institute. Many of the proposed undertakings I cannot really judge for myself. I have to depend on the reports of my colleagues, and this I can do, because I have chosen reliable collaborators. Beyond that, however, I still have two criteria on which to base my decision when, as is often the case, I know too little about the matter. The first criterion is the scientific success of the division in question. As a rule, I shall allot a division a larger share of the institutional budget if on the average it has been particularly successful over a considerable period, and a smaller share if the work seems less promising. The second criterion is the comparison with other effectively operating institutes abroad. If I find, for example, that in comparable research institutes abroad three technicians are generally allotted to one scientist, I shall assume, if I hear of no very cogent reasons against it, that in our institute we ought to aim for the same ratio.

I have already mentioned that in other industrial countries, such as Britain, France and the United States, the public expenditure on scientific research and development seems to be about twice that in the Federal Republic. Naturally, in theory it could be that the others are doing things wrong and that we alone are doing them right, but results over the last thirty years do not support that view; and before this the differences were certainly not so great. An objection sometimes heard in such discussions is this: "In order to prepare our plans in the Federal budget, we cannot start from such comparative figures, but must have precise amounts in Deutschmarks. If the tax yield rises, and with it the state budget, then we can take on additional tasks, but otherwise we must simply limit them." At bottom, this objection poses this question: "How are the comparative figures to alter, supposing the overall Federal budget does rise or fall?" If such a change in the numerical size of the Federal budget is due merely to changes in the value of money, i.e., in its purchasing power, this ought naturally to have no influence at all on the disposition of the Federal budget. If we are dealing, however, with a real change in its value—if owing, say, to a decline in industrial production the total tax yield should actually diminish in value, though perhaps not numerically—then this simple answer cannot be given. One may then, perhaps, draw a comparison with a familiar situation in agriculture. If the harvest has been bad, will the farmer raise or lower the proportion of his crop that he customarily puts aside as seed corn for the following year? I assume that he will raise it, although he can then bake less bread or sell less seed corn. But at least he will surely aim for a better harvest next year. In this analogy, scientific research is the seed corn from which future economic welfare, higher revenue, efficient running of the country and much else are due to spring, and in many places have indeed sprung in the past.

Thus a partial answer to our original question might run: In the past decade we have spent on scientific research only about half of the public funds that have been expended in comparable areas abroad, and hence we have also achieved correspondingly less. This answer is probably rather too simple. I don't believe the scientific achievements of a people to be so directly proportional to the financial means employed. That is evident even from the example of Germany after the First World War. But indirectly the connection may still exist. The proportion of the national budget applied to the furthering of research is an index of the importance attaching to scientific research in the public mind. This public estimation of science is probably a very important incentive for the younger generation to achieve something in science, and during the twenties it was very high.

I have mentioned that the style of research has changed and that our administration has had great difficulties in adapting itself to this style. I should like here to pick out two problems that seem to me particularly characteristic: the organization of the major institutes and the freedom of movement of scientists. Major institutes, also often rather disparagingly called "research factories," are institutes in which a specific research goal is pursued by the use of very large and costly technical means.

They arose in this form in various countries only after the Second World War, and have to be financed from public funds, since they bring in no immediate economic return. In the Federal Republic, I may mention as examples the reactor stations at Karlsruhe and Jülich, the big Desy accelerator in Hamburg, and the Institute for Plasma Physics at Munich-Garching. A really satisfactory legal form for these institutes has yet to be found. The earlier patterns, especially the form of organization of the Federal institute, are not particularly suitable here; not even where the Federal Government is the sole source of finance. The Federal institute pattern is appropriate where there is a steady supply of routine scientific work that can be carried out by careful reliable officials. But it does not suit where new paths in science have continually to be trodden, where the keen wind of international competition enforces a constant adaptation to new scientific findings and methods, where care must be taken, even through a frequent interchange of scientists with research institutes abroad, to see that we constantly keep working at the outermost frontier of science. I think it of great importance that we push forward here to new forms of organization. Perhaps use can be made of the negotiations now pending over the institute in Munich-Garching to search out new and better methods here. The leading positions in such institutes must be occupied by scientists of international standing—but this means scientists who are also subject to good offers from abroad. And so here there is no avoiding a certain adjustment to conditions overseas, in America, in Euratom, etc. We can no longer employ without question our former German patterns of administration, because then we shall no longer be able to obtain the best-qualified people.

For example, we have hitherto been accustomed to conclude salary contracts with scientists, which begin at a relatively modest initial remuneration, allow for increases every two years and provide good security for long periods, often to an advanced age. In America, contracts are usually made for three to five years only, and there is no talk of increases—instead, the starting salary is at least 50 percent higher. The abler men will gladly take the risk of the three-year contract; they expect to be able to achieve so much in three years that they will be reappointed. The less capable will prefer the longer-term security. Thus our salary system ensures that the ablest people are constantly departing for America, while the less efficient remain at home. However, it is by no means just a matter of the material status, i.e., the income of the scientist in question. Equal importance attaches to responsible participation in the latest research, or the chance of a certain freedom in the choice of collaborators, in cooperating with overseas institutes, in travel to such institutes or congresses, etc. We are constantly finding that the old and formerly trustworthy patterns of administration constrict life too much here, and that young Germans who have worked for a spell in America are no longer willing—precisely for fear of this constriction—to return to Germany. Here, therefore, we must adapt our patterns of administration to those of the outside world. The rigid application of existing administrative principles would render impossible a modern style of scientific research work, one adapted to international life.

It would probably bring about an improvement in the situation if we were to adopt a custom that has long been a matter of course in the Anglo-Saxon countries. It is common there for men who have been brought up and gained experience in science, technology or economics to enter government administration later on, while men who have been employed in government service may move over into responsible administrative positions in research institutes. A similar flexibility would be highly desirable here in the Federal Republic. For a time, therefore, there will probably have to be some experimenting in the administrative field. Rigidity would be the chief danger here. We have difficulties with scientific research in Germany because our old patterns of administration are no longer suited to the new style of science.

As the Federal Chancellor stressed in his official statement, in matters of contested jurisdiction arrangements must be made between the Länder and the Federal Government, which will regulate authority in a clear, realistic and rational manner; so there is no need to say more on that subject here. In the past, however, there have actually been occasional difficulties precisely where the Federal Government constitutionally carries the supreme responsibility, namely in this very field of scientific research. I am thinking especially in this connection of the oft-discussed Clause 950 in the Federal budget, "Promotion of Atomic Research through Appropriations for the Modernizing and Extension of Scientific Institutes and Establishments." Atomic research in the technical colleges largely depends on this clause. In the past year this appropriation was sharply curtailed at a relatively late date, and this meant, it was anticipated, that in virtue of their sovereignty in cultural matters the Länder would step in here. What has happened can be most simply explained by the example of the Max Planck Institute for Physics and Astrophysics in Munich, though in many institutes of higher technical learning the results were even worse.

Owing to this ordinance, the funds already allocated to our Institute were suddenly cut by the very large amount of 700,000 Deutschmarks. On the very same day that I received this news, I naturally applied to the Bavarian Ministry of Culture to see whether it could make the missing funds available to us. The answer, as might be expected, was that this was unfortunately impossible, since the funds of the Ministry were already completely committed. So that is not the way to do it. If you consider a measure to be necessary and believe that the Land can carry it out, you must first make certain that it will also actually be carried out by the Land. If you don't consider it necessary, then you must say so clearly at the earliest possible moment, for of course it is impossible to direct even an institute sensibly unless one has a reasonably exact idea of the funds at one's disposal.

I should now like to refer briefly to certain special fields of scientific research, which occupy a peculiar position, because in them gigantic efforts have been made with a view to great-power defense and because international involvement here is very much closer than in other fields. I am thinking especially of atomic and space research. The great powers are investing immense sums in these areas because—rightly or wrongly—they are fearful of falling behind in the technological race that forms the basis of any armament program. For smaller countries, such as the Federal Republic, this latter reason is initially of no account, since defense is primarily the business of the great powers, who enlist their allies in its burdens only to a limited degree. But there are two pertinent reasons why industrial countries like the Federal Republic must take an energetic part in such work. The enormous technical efforts of the great powers, which are financed by the public purse, are effecting a constant enlargement of technical knowledge; new materials are being developed, new methods devised, new technical possibilities discovered. Think, for example, of the development of miniature electronic devices, computers, guidance systems and so on in connection with rocket research. A country that takes no part in the work going on in these areas will remain permanently backward from a technical point of view. Let me remind you of the example from the Munich Institute that I mentioned right at the outset: in the long run such backwardness may have serious consequences for the economy.

Such work, to be sure, is initially unrelated to any immediate economic advantage, but indirectly the well-known fable of the orchard may apply. A father, on his deathbed, bequeaths an orchard to his sons and tells them there is a great treasure lying hidden there, which they are to dig for after his death. This they do, but despite the most careful and repeated excavations, they are disappointed to find nothing. In the following summer, however, the orchard bears more fruit than ever before, and they gradually realize what treasure it was that their father meant. This indirect benefit of major technical efforts can thus be extremely important.

The second and perhaps still more pertinent reason consists in the incentive to international collaboration. Precisely because of the enormous resources that have to be employed in them, these areas of research are being cultivated by many countries working in common. The Federal Republic has been invited to take part in such international projects as the research programs of Euratom, CERN, ESRO and ELDO, i.e., for the exploitation of atomic energy, high-energy physics, space and rocket research; and during the past ten years we have actually collaborated to some extent, for example, in the international CERN organization in Geneva. If it is asked how far the Federal Republic should participate in such undertakings, we must be clear from the outset that a limit will unavoidably be set by the funds at our disposal. On the one hand, the contributions to international organizations are already considerable, while on the other, participation makes sense at all only if one is ready to prosecute the research in question with great vigor in one's own country, that is, to devote considerably larger sums still to home-based research of this kind than to the international organizations. For the international membership contribution would be completely wasted if it did not lead to a fruitful development of the relevant research area at home, and this it can naturally do only if great efforts are made there. Thus the Federal Republic has hitherto derived too little benefit from its participation in the CERN organization, because German efforts in the field of high-energy physics have been too small. We hope that when the big Desy accelerator in Hamburg comes into operation, this situation will soon improve.

In view of the high and rising costs entailed by every such collaboration in international projects, especially at home, we shall thus have to weigh very cautiously where we are going to join in. It will certainly not be possible to take part in everything. The decision about which projects are to be selected, like that over points of concentration, can only be arrived at after careful consultation in commissions made up of representatives from science and technology and the ministries involved. Even the international negotiations can scarcely be satisfactorily conducted without the participation of the scientific specialists. But where we do take part, we should do so in full force, and thus also make great efforts at home. Nothing is more uncongenial, even in the international field, than a collaboration agreed to but granted only in a halfhearted fashion.

Concerning these vast and extraordinarily costly international projects, which are really enterprises of mankind, many will ask whether they are absolutely necessary, whether the huge sums involved could not better be employed in some other way. I think it is important to remember here that we in the Federal Republic have very little say in the matter. These great projects will be carried out regardless of whether we join in or not. We merely have the choice of coming in or staying out. We are therefore in somewhat the same position as a schoolboy whose class has decided to undertake a hiking trip in the summer to Scandinavia. They are all collecting cash, contributing equipment, tents and rucksacks, looking forward to the trip. But the schoolboy does not know whether to take part. He jibs at the expense, Scandinavia does not interest him particularly, nor have his relations with his schoolfellows always been of the best. Should he join in? I think, at all events, that he ought not to be content with contributing twenty Deutschmarks to the travel fund as a sign of goodwill. That would be pointless. It would probably be better to join the party, to venture forth with them into the unknown and to enjoy the novelty of the thing.

To apply this to our present situation: If we join in, if we can take a common satisfaction in the success of these great human enterprises, we shall quite certainly find it easier to retain the most gifted of our young scientists here in Germany and to reap the benefit of their talents. I venture to think that this would in fact hold good for a much larger group than just that of the younger generation of scientists. To participate in great enterprises, even if it costs much trouble and effort to realize them and even if one may not be altogether certain of their value, is for most people more satisfying than mere material well-being and comfort. I cannot share the pessimistic view of our countrymen which is often stated by saying that the only road to success at election time is to promise less work, more leisure and comfort, and higher pay. Our people are not like that. Perhaps, indeed, the hearts of men will sooner be won by setting exalted goals and showing a real willingness to take part in the common creation of this extraordinary modern world. For only those who take part can also influence the course of that world in the direction they think desirable.

IX

Natural Law and the Structure of Matter

It was here in this part of the world, on the coast of the Aegean Sea, that the philosophers Leucippus and Democritus pondered about the structure of matter, and down there in the market-place, where twilight is now falling, that Socrates disputed about the basic difficulties in our modes of expression and Plato taught that the Idea, the form, was the truly fundamental pattern behind the phenomena. The problems first formulated in this country two and a half thousand years ago have occupied the human mind almost unceasingly ever since, and have been discussed again and again in the course of history whenever new developments have altered the light in which the old lines of thought appeared.

If I endeavor today to take up some of the old problems concerning the structure of matter and the concept of natural law, it is because the development of atomic physics in our own day has radically altered our whole outlook on nature and the structure of matter. It is perhaps not an improper exaggeration to maintain that some of the old problems have quite recently found a clear and final solution. So it is permissible today to speak about this new and perhaps conclusive answer to questions that were formulated here thousands of years ago.

Address delivered on the hill of Pnyx, opposite the Acropolis in Athens, on June 3, 1964. First published in a bibliophile edition (German and English) in the Belser-Presse collection *Meilensteine des Denkens und Forschens,* Stuttgart 1967. English text reissued, 1970, by Rebel Press, London. [The present translation is based upon the latter.—Tr.]

There is, however, yet another reason for renewing consideration of these problems. The philosophy of materialism, developed in antiquity by Leucippus and Democritus, has been the subject of many discussions since the rise of modern science in the seventeenth century and, in the new form of dialectical materialism, has been one of the moving forces in the political changes of the nineteenth and twentieth centuries. If philosophical ideas about the structure of matter have been able to play such a role in human life, if in European society they have operated almost like an explosive and may yet perhaps do so in other parts of the world, it is even more important to know what our present scientific knowledge has to say about this philosophy. To put it in rather more general and precise terms, we may hope that a philosophical analysis of recent scientific developments will contribute to a replacement of conflicting dogmatic opinions about the basic problems we have broached, by a sober readjustment to a new situation, which in itself can even now be regarded as a revolution in human life on this earth. But even aside from this influence of science upon our time, it may be of interest to compare the philosophical discussions in ancient Greece with the findings of experimental science and modern atomic physics. If I may already anticipate at this point the outcome of such a comparison; it seems that, in spite of the tremendous success that the concept of the atom has achieved in modern science, Plato was very much nearer to the truth about the structure of matter than Leucippus or Democritus. But it will doubtless be necessary to begin by repeating some of the most important arguments adduced in the ancient discussions about matter and life, being and becoming, before we can enter into the findings of modern science.

1. *The Concept of Matter in Ancient Philosophy*

At the beginning of Greek philosophy there stood the dilemma of the "one" and the "many." We know that there is an ever-changing variety of phenomena appearing to our senses. Yet we believe that ultimately it should be possible to trace them back somehow to some one principle. We try, of course, to understand the phenomena, and in so doing we realize that all understanding begins with perceiving similarities and regularities among them. The regularities are then apprehended as special consequences of something that is common to the different phenomena, and may therefore be called an underlying principle. In this way, every attempt to understand the changing multiplicity of the phenomena must become a search for an underlying principle. It was a characteristic feature of the way of thinking in ancient Greece that the first philosophers looked for a "material cause" of all things. That seems at first to be a very natural starting point for a world which certainly consists of matter. But from there we also lapse at once into a dilemma, namely whether this material cause of everything should be identified with one of the existing forms of matter, such as "water" in the philosophy of Thales or "fire" in the teaching of Heraclitus, or whether some fundamental substance should be assumed, of which real matter presents only the transient forms. These two possibilities were worked out in ancient philosophy, but we shall not discuss them in detail here.

If one pursues such a line of thought, the underlying principle, the hope for simplicity in the phenomena, becomes associated with a "fundamental substance." This question then arises: At what point or in what way can the simplicity in the behavior of the basic substance be formulated? For this simplicity is not immediately discernible in the phenomena. Water can turn into ice or make flowers spring from the soil. But the smallest parts of water, which perhaps are the same in ice or steam or in the flowers, could be the simple element. Their behavior could be determined by simple laws, and these laws could then be formulated.

In this way the concept of the "smallest parts of matter" is a natural consequence of the quest for simplicity, once attention is primarily directed to matter, to the material cause of everything.

On the other hand, this concept of the smallest parts of

matter, whose laws are to be simply understood, leads at once into the well-known difficulties connected with the concept of infinity. A piece of matter can be divided, the parts can be separated into smaller pieces still, these pieces yet again be divided, and so on; but for all that, it is hard for us to imagine that this divisibility should continue to infinity. It seems to us somehow more natural to assume that there are smallest parts that cannot be divided further. On the other hand, we are also unable to imagine it impossible in principle to subdivide these smallest parts still further. At least we can always picture still smaller parts in thought, we can suppose that on a much smaller scale we encounter the same situation as on the ordinary scale. Thus we are manifestly led astray by our own powers of imagination when we attempt to visualize the process of ever-continuing division. This was also the feeling of the Greek philosophers, and the "atomic hypothesis," the idea of smallest indivisible parts, can be regarded as a first and natural way out of the difficulty.

The founders of atomism, Leucippus and Democritus, tried to avoid the difficulty by assuming the atom to be eternal and indestructible, the only thing really existing. All other things exist only because they are composed of atoms. The antithesis of "being" and "non-being" in the philosophy of Parmenides is here coarsened into that between the "full" and the "void." Being is not only one; it can be repeated infinitely many times. Being is indestructible, and therefore the atom, too, is indestructible. The void, the empty space between the atoms, allows for position and motion, and thus for properties of the atom, whereas by definition, as it were, pure being can have no other property than that of existence.

This latter part of the doctrine of Leucippus and Democritus is at once its strength and its weakness. On the one hand it provides an immediate explanation of the different aggregate states of matter, such as ice, water and steam, since the atoms may lie densely packed and in order beside each other, or be caught in disorder and irregular motion, or finally be separated at fairly large relative intervals in space. This part of the atomic hypothesis was therefore to prove exceedingly fruitful at a later stage. On the other hand, the atom becomes in this fashion a mere building block of matter; its properties, position and motion in space turn it into something quite different from what was meant by the original concept of "being." The atoms can even have a finite extension, and here we have finally lost the only convincing argument for their indivisibility. If the atom has spatial properties, why should it not be divided? At least its indivisibility then becomes a physical, not a fundamental property. We can now again ask questions about the structure of the atom, and we run the risk of losing all the simplicity we had hoped to find among the smallest parts of matter. We get the impression, therefore, that in its original form the atomic hypothesis was not sufficiently subtle to explain what the philosophers really wished to understand: the simple element in the phenomena and in the structure of matter.

Still, the atomic hypothesis does go a large part of the way in the right direction. The whole multiplicity of diverse phenomena, the many observed properties of matter, can be reduced to the position and motion of the atoms. Properties such as smell or color or taste are not present in atoms. But their position and motion can evoke these properties indirectly. Position and motion seem to be much simpler concepts than the empirical qualities of taste, smell or color. But then it naturally remains to ask what determines the position and motion of the atoms. The Greek philosophers did not attempt at this point to formulate a law of nature; the modern concept of natural law did not fit into their way of thought. Yet they seem to have thought of some kind of causal description, or determinism, since they spoke of necessity, of cause and effect.

The intention of the atomic hypothesis had been to point the way from the "many" to the "one," to formulate the underlying principle, the material cause, by virtue of which all phenomena can be understood. The atoms could be regarded as the material cause, but only a general law determining their positions and velocities could actually play the part of the fundamental principle. However, when the Greek philosophers discussed the laws of nature, their thoughts were directed to static forms, geometrical symmetries, rather than to processes in space and time. The circular orbits of the planets, the regular geometrical solids, appeared to be the permanent structures of the world. The modern idea, that the position and velocity of the atom at a given time could be uniquely connected by a mathematical law with its position and velocity at a later time, did not fit into the pattern of thought of that era, since it employs the concept of time in a manner that arose only out of the thinking of a much later epoch.

When Plato himself took up the problems raised by Leucippus and Democritus, he adopted the idea of smallest units of matter, but he took the strongest exception to the tendency of that philosophy to suppose the atoms to be the foundation of all existence, the only truly existing material objects. Plato's atoms were not strictly material, being thought of as geometrical forms, the regular solids of the mathematicians. These bodies, in keeping with the starting point of his idealistic philosophy, were in some sense the Ideas underlying the structure of matter and characterizing the physical behavior of the elements to which they belonged. The cube, for example, was the smallest particle of the element earth, and thereby symbolized at the same time the earth's stability. The tetrahedron, with its sharp points, represented the smallest particle of the element fire. The icosahedron, which comes closest among the regular solids to a sphere, stood for the mobility of the element water. In this way the regular solids were able to serve as symbols for certain tendencies in the physical behavior of matter.

But they were not strictly atoms, not indivisible basic units like those of the materialist philosophy. Plato regarded them as composed from the triangles forming their surfaces; and therefore, by exchanging triangles, these smallest particles could be commuted into each other. Thus two atoms of air, for example, and one of fire could be compounded into an atom of water. In this way Plato was able to escape the problem of the infinite divisibility of matter. For as two-dimensional surfaces the triangles were not bodies, not matter any longer; hence matter could not be further divided ad infinitum. At the lower end, therefore, in the realm, that is, of minimal spatial dimensions, the concept of matter is resolved into that of mathematical form. This form determines the behavior, first of the smallest parts of matter, then of matter itself. To a certain extent it replaces the natural law of later physics; for without making explicit reference to the course of time, it characterizes the tendencies in the behavior of matter. One might say, perhaps, that the fundamental tendencies were represented by the geometrical shape of the smallest units, while the finer details of those tendencies found expression in the relative position and velocity of these units.

This whole description fits exactly into the central ideas of Plato's idealist philosophy. The structure underlying the phenomena is not given by material objects like the atoms of Democritus but by the form that determines the material objects. The Ideas are more fundamental than the objects. And since the smallest parts of matter have to be the objects whereby the simplicity of the world becomes visible, whereby we approximate to the "one," the "unity" of the world, the Ideas can be described mathematically—they are simply mathematical forms. The saying "God is a mathematician," which in this form assuredly derives from a later period of philosophy, has its origin in this passage from the Platonic philosophy.

The importance of this step in philosophical thought can hardly be reckoned too highly. It can be seen as the decisive beginning of the mathematical science of nature, and hence be made responsible also for the later technical applications that have altered the whole picture of the world. By this step it is also first established what the term "understanding" is to mean. Among all the possible forms of understanding, the one form practiced in mathematics is singled out as the "true" understanding. Whereas all language, indeed, all art and all poetry in some way mediate understanding, it is here maintained that only the employment of a precise, logically consistent language, a language so far capable of formalization that proofs become possible, can lead to true understanding. One feels the strength of the impression made upon the Greek philosophers by the persuasive force of logical and mathematical arguments. They are obviously overwhelmed by this force. But perhaps they surrendered too early at this point.

2. *The Answer of Modern Science to the Old Problems*

The most important difference between modern science and ancient natural philosophy lies in the method employed. Whereas in ancient philosophy the empirical knowledge of natural phenomena was reckoned sufficient for drawing conclusions about the underlying principles, it is a characteristic feature of modern science to institute experiments, i.e., to put specific questions to nature, whose answers are then to yield information about its laws. This difference of method also leads in the sequel to a very different way of looking at things. Attention is directed not so much to the fundamental laws as to the regularities among the details. Natural science is developed, so to speak, from the other end, not from the general laws but from individual groups of phenomena, in which nature has answered the questions put by experiment. Since the time when Galileo is fabled to have dropped his stones from the leaning tower of Pisa, in order to study the laws of falling bodies, science has been busy with the details of the most diverse phenomena, with falling stones, with the moon's motion about the Earth, with waves in water, with light rays refracted through a prism, etc. Even when Isaac Newton, in his great work *Principia Mathematica,* had elucidated the various mechanical processes by means of a unitary law, attention was concentrated upon the details that were to be derived from the underlying mathematical principles. A result that was correct, i.e., in accord with experience, in deriving the details was reckoned as the decisive criterion for the correctness of the theory.

This change in the whole way of looking at things also had other important consequences. An exact knowledge of details can be useful in practice. Within certain limits it puts man in a position to direct phenomena according to his will. The technical applications of modern science therefore begin with the knowledge of details. In this way the concept of "natural law" also gradually alters in meaning; its main significance no longer lies in its generality but in its consequences with regard to details. The law becomes a prescription for technical applications. The most important feature of natural law is now held to be that of making it possible to predict the outcome of a given experiment.

It is easy to see that in such a natural science the concept of time must play a part entirely different from that in ancient philosophy. It is not an eternal, unvarying structure that is set forth in a law of nature; it is a matter, rather, or regularity in changes over time. If a natural law of this type is formulated in an exact mathematical language, the physicist is at once presented with an infinite variety of experiments he could perform in order to test the correctness of the proposed law, while a single disagreement between theory and experiment could refute the theory. This situation lends enormous weight to the mathematical formulation of a natural law. If all known experimental facts agree with the results mathematically derived from the law, it becomes extraordinarily difficult to doubt its general validity. It is therefore intelligible that Newton's *Principia* should have dominated science for more than two centuries.

If we trace the history of physics from Newton to the present day, we see that, despite the interest in details, very general laws of nature have been formulated on several occasions. The nineteenth century saw an exact working out of the statistical theory of heat. The theories of electromagnetism and special relativity have proved susceptible of combination into a very general group of natural laws containing statements not only about electrical phenomena but also about the structure of space and time. In our own century, the mathematical formulation of the quantum theory has led to an understanding of the outer shells of chemical atoms, and thus of the chemical properties of matter generally. The relations and connections between these different laws, especially between relativity and quantum theory, are not yet fully explained. But the latest developments in particle physics permit one to hope that these relations may be satisfactorily analyzed in the relatively near future. We are thus already in a position to consider what answers can be given by this whole scientific development to the questions of the old philosophers.

During the nineteenth century, the development of chemistry and the theory of heat conformed very closely to the ideas first put forward by Leucippus and Democritus. A revival of the materialist philosophy in its modern form, that of dialectical materialism, was thus a natural counterpart to the impressive advances made during this period in chemistry and physics. The concept of the atom had proved exceptionally fruitful in the explanation of chemical bonding and the physical behavior of gases. It was soon found, however, that the particles called atoms by the chemists were composed of still smaller units. But these smaller units, the electrons, followed by the atomic nuclei and finally the elementary particles, protons and neutrons, also still seemed to be atoms from the standpoint of the materialist philosophy. The fact that, at least indirectly, one can actually see a single elementary particle—in a cloud chamber, say, or a bubble chamber—supports the view that the smallest units of matter are real physical objects, existing in the same sense that stones or flowers do.

But the inherent difficulties of the materialist theory of the atom, which had become apparent even in the ancient discussions about smallest particles, have also appeared very clearly in the development of physics during the present century. First of all, there is the problem of the infinite divisibility of matter.

The so-called atoms of the chemists had proved to be made up of nuclei and electrons. The atomic nucleus had been split up into protons and neutrons. Is it not possible, we are bound to ask, to go on dividing even the elementary particles? If the answer to this question is "yes," then even the elementary particles are not atoms in the Greek sense, not indivisible units. If it is "no," we must explain why the elementary particles cannot be further divided. Hitherto, indeed, it had always been possible, sooner or later, to split even those particles which had long been regarded as the smallest units, once sufficiently large forces were employed to do this. It was therefore natural to assume that by increasing the forces, i.e., by simply increasing the energy in the collision of the particles, one should finally be able to disrupt even protons and neutrons. And this would probably mean that we never do come to an end, that there simply are no smallest units of matter. But before entering upon a discussion of the present-day solution to this problem, I must mention the second difficulty.

This second difficulty relates to the question whether the smallest units are ordinary physical objects, whether they exist in the same way as stones or flowers. Here, the development of quantum theory some forty years ago has created a complete change in the situation. The mathematically formulated laws of quantum theory show clearly that our ordinary intuitive concepts cannot be unambiguously applied to the smallest particles. All the words or concepts we use to describe ordinary physical objects, such as position, velocity, color, size and so on, become indefinite and problematic if we try to use them of elementary particles. I cannot enter here into the details of this problem, which has been discussed so frequently in recent years. But it is important to realize that while the behavior of the smallest particles cannot be unambiguously described in ordinary language, the language of mathematics is still adequate for a clear-cut account of what is going on.

The latest advances in the field of particle physics have also, however, provided a solution to the problem mentioned first, the riddle of the infinite divisibility of matter. During the postwar period, large accelerators have been built in different parts of the world, in order, if possible, to split even the elementary particles further still. The results seem highly surprising to anyone who has not yet learned that our ordinary concepts are not appropriate to the smallest units of matter. If two elementary particles collide with extremely high energy, they actually fall to pieces, as a rule, and often into many pieces, but the pieces are no smaller than the particles that were split. Independently of the energy available (if only this is large enough), we always obtain from such a collision the same sorts of particle that we have been familiar with for a number of years. Even in cosmic radiation, where the available energy of a particle can in some cases be a thousand times larger than in the biggest existing accelerator, no other or smaller particles have been found. Their charge, for example, can easily be measured, and it is always equal to, or an integer multiple of, the charge of the electron.

The best description of these collision phenomena is therefore not to assert that the colliding particles have been broken up but to speak of the emergence of new particles from the collision energy, in accordance with the laws of the theory of relativity. We can say that all particles are made of the same fundamental substance, which can be designated energy or matter; or we can put things as follows: the basic substance "energy" becomes "matter" by assuming the form of an elementary particle. In this way the new experiments have taught us that we can combine the two seemingly conflicting statements: "Matter is infinitely divisible" and "There are smallest units of matter," without running into logical difficulties. This surprising result again underlines the fact that our ordinary concepts cannot be applied unambiguously to these smallest units.

During the coming years, the high-energy accelerators will bring to light many further interesting details about the behavior of elementary particles. But I am inclined to think that the answer just considered to the old philosophical problems will turn out to be final. If this is so, does this answer confirm the views of Democritus or Plato?

I think that on this point modern physics has definitely decided for Plato. For the smallest units of matter are in fact not physical objects in the ordinary sense of the word; they are forms, structures or—in Plato's sense—Ideas, which can be unambiguously spoken of only in the language of mathematics. Democritus and Plato both had hoped that in the smallest units of matter they would be approaching the "one," the unitary principle that governs the course of the world. Plato was convinced that this principle can be expressed and understood only in mathematical form. The central problem of theoretical physics nowadays is the mathematical formulation of the natural law underlying the behavior of the elementary particles. From the experimental situation we infer that a satisfactory theory of the elementary particles must at the same time be a theory of physics in general; and hence of everything else belonging to this physics.

In this way a program could be carried out that in modern times was first proposed by Einstein: a unified theory of matter—and hence simultaneously a quantum theory of matter—could be formulated, which might serve quite generally as a foundation for physics. We do not yet know whether the mathematical forms proposed for this unifying principle are already adequate or will have to be replaced by forms more abstract still. But our present knowledge of the elementary particles is certainly enough for us to say what the main content of this law has to be. It must essentially set forth a small number of fundamental symmetry properties in nature, which have been known empirically for some years; and in addition to these symmetries, it must contain the principle of causality as understood in relativity theory. The most important of the symmetries are the so-called "Lorentz group" of special relativity theory, which includes the key statements about space and time, and the so-called "isospin group," which has to do with the electric charge on the elementary particles. There are also other symmetries, but of these I shall say nothing here. Relativistic causality is connected with the Lorentz group but must be considered an independent principle.

This situation reminds us at once of the symmetrical bodies introduced by Plato to represent the fundamental structures of matter. Plato's symmetries were not yet the correct ones; but he was right in believing that ultimately, at the heart of nature, among the smallest units of matter, we find mathematical symmetries. It was an unbelievable achievement of the ancient philosophers to have asked the right questions. But, lacking all knowledge of the empirical details, we could not have expected them to find answers that were correct in detail as well.

3. *Consequences for the Evolution of Human Thought in Our Own Day*

The search for the "one," for the ultimate source of all understanding, has doubtless played a similar role in the origin of both religion and science. But the scientific method that was developed in the sixteenth and seventeenth centuries, the interest in those details which can be tested by experiment, has for a long time pointed science along a different path. It is not surprising that this attitude should have led to a conflict between science and religion, as soon as a law contradicted, in some particular and perhaps very important detail, the general picture, the mode and manner, in which the facts had been spoken of in religion. This conflict, which began in modern times with the celebrated trial of Galileo, has been discussed often enough, and I need not repeat this discussion here. One may recall that even in ancient Greece Socrates was condemned to death because his teachings seemed to contradict the traditional religion. In the nineteenth century, this conflict reached its peak in the attempt of some philosophers to replace traditional Christianity by a scientific philosophy, based upon a materialist version of the Hegelian dialectic. It might be said that in directing their gaze upon a materialist interpretation of the "one," the scientists were attempting to find their way back again to this "one" from the multitude of details.

But here too the split between the "one" and the "many" is not easily overcome. It is no mere accident that in certain countries, where dialectical materialism has been proclaimed in our century as the official creed, the conflict between science and the accepted doctrines has not been altogether possible to avoid. A particular scientific finding, the result of new observations, can seemingly run counter to the official view. If it is true that the harmony in a community is provided by its relation to the "one"—in whatever terms the "one" may actually be referred to—then it is easy to see that an apparent contradiction between an accepted scientific finding and the received habits of talk about the "one" becomes a serious problem. The history of recent years contains numerous examples of political difficulties that have arisen on this score. We learn from this that it is not primarily a matter of the struggle between two conflicting theories, such as materialism and idealism, but of a quarrel between the scientific method, namely the ascertaining of particularity on the one hand, and the common relation to the "one" on the other. The great success of the scientific method, by trial and error, excludes in our own day any definition of truth that has not withstood the severe criteria of this method. But at the same time it also seems to be a well-established finding of social science that the internal equilibrium of a society depends, to some extent at least, on a common relation to the "one." Hence the search for the "one" can scarcely be forgotten.

If modern science has something to contribute to this problem, it is not by deciding for or against one of these doctrines; for example, as was possibly believed in the nineteenth century, by coming down in favor of materialism and against the Christian philosophy, or, as I now believe, in favor of Plato's idealism and against the materialism of Democritus. On the contrary, the chief profit we can derive in these problems from the progress of modern science is to learn how cautious we have to be with language and with the meaning of words. I would therefore like to devote the last part of my address to a few remarks about the problem of language in modern science and ancient philosophy.

If we may take our cue at this point from Plato's dialogues, the unavoidable limitations of our means of expression were already a central theme in the philosophy of Socrates; one might even say that his whole life was a constant battle with these limitations. Socrates never wearied of explaining to his countrymen, here on the streets of Athens, that they did not know exactly what they meant by the words they were employing. The story goes that one of Socrates' opponents, a sophist who was annoyed at Socrates' constant reference to this insufficiency of language, criticized him and said: "But Socrates, this is a bore; you are always saying the same about the same." Socrates replied: "But you sophists, who are so clever, perhaps never say the same about the same."

The reason for laying such stress on this problem of language was doubtless that Socrates was aware, on the one hand, of how many misunderstandings can be engendered by a careless use of language, how important it is to use precise terms and to elucidate concepts before employing them. On the other hand, he probably also realized that this may ultimately be an insoluble task. The situation confronting us in our attempt to "understand" may drive us to conclude that our existing means of expression do not allow of a clear and unambiguous description of the facts.

The tension between the demand for complete clarity and the inevitable inadequacy of existing concepts has been especially marked in modern science. In atomic physics we make use of a highly developed mathematical language that satisfies all the requirements in regard to clarity and precision. At the same time we recognize that we cannot describe atomic phenomena without ambiguity in any ordinary language; we cannot, for example, speak unambiguously about the behavior of an electron in the interior of an atom. It would be premature, however, to insist that we should avoid the difficulty by confining ourselves to the use of mathematical language. This is no genuine way out, since we do not know how far the mathematical language can be applied to the phenomena. In the last resort, even science must rely upon ordinary language, since it is the only language in which we can be sure of really grasping the phenomena.

This situation throws some light on the tension between the scientific method, on the one hand, and the relation of society to the "one," the fundamental principle behind the phenomena, on the other. It seems obvious that this latter relation cannot and should not be expressed in a precise and highly sophisticated language whose applicability to the real world may be very restricted. The only thing that will do for this purpose is the natural language everyone can understand. Reliable results in science, however, can be secured only by unambiguous statement; here we cannot do without the precision and clarity of an abstract mathematical language.

This necessity of constantly shuttling between the two languages is unfortunately a chronic source of misunderstandings, since in many cases the same words are employed in both. The difficulty is unavoidable. But it may yet be of some help always to bear in mind that modern science is obliged to make use of both languages, that the same word may have very different meanings in each of them, that different criteria of truth apply and that one should not therefore talk too hastily of contradictions.

If we wish to approach the "one" in the terms of a precise scientific language, we must turn our attention to that center of science described by Plato, in which the fundamental mathematical symmetries are to be found. In the concepts of this language we must be content with the statement that "God is a mathematician"; for we have freely chosen to confine our vision to that realm of being which can be understood in the mathematical sense of the word "understanding," which can be described in rational terms.

Plato himself was not content with this restriction. Having pointed out with the utmost clarity the possibilities and limitations of precise language, he switched to the language of poetry, which evokes in the hearer images conveying understanding of an altogether different kind. I shall not seek to discuss here what this kind of understanding can really mean. These images are probably connected with the unconscious mental patterns the psychologists speak of as archetypes, forms of strongly emotional character that in some way reflect the internal structures of the world. But whatever the explanation for these other forms of understanding, the language of images and likenesses is probably the only way of approaching the "one" from more general domains. If the harmony in a society rests on a common interpretation of the "one," the unitary principle behind the phenomena, then the language of poetry may be more important here than the language of science.

X

Goethe's View of Nature and the World of Science and Technology

Goethe's view of nature and the world of science and technology is a topic as old as Goethe's own efforts to understand nature, as his own practice of natural science; for Goethe lived during the beginnings of that scientific and technical world which is all about us today. Much has been said on this matter by Goethe himself, by his contemporaries and by scientists and philosophers since. We have long known what an important role the issue played in Goethe's life, and we also know how much of our contemporary world is called into question when we measure our scientific and technical achievements against Goethe's demands. It has often been pointed out, moreover, how touchily Goethe reacted to the cleavage between his theory of colors and the generally accepted optics of Newton, how violent and irrelevant his polemic against Newton was wont to be; and it has also been noted that his critique of romanticism, his fundamentally negative attitude to romantic art, displays a certain inner relationship to his polemic against the prevailing natural science. So much has been said and written on all this, and the underlying problems have been so thoroughly illuminated from so many sides, that there is scarcely anything left to

Lecture to the Plenary Session of the Goethe Society in Weimar, on May 21, 1967. First published in *Goethe*—New Series of the *Goethe Society Yearbook*, ed. Andreas B. Wachsmuth, Vol. 29. Weimar (Hermann Bohlaus Nachfolger) 1967, pp. 27–42.

do but to follow out the oft-expressed ideas a little further, and to examine them in the light of an acquaintance with the present-day scientific and technical world, especially with the latest developments in natural science.

In so doing, we shall not allow ourselves to be guided in advance by the pessimistic viewpoint, echoed, for example, in Karl Jaspers, that because Goethe shut himself off before the coming technical world, because he did not acknowledge the task of finding man's way in this new world, he has nothing more to tell us on this subject today. On the contrary, we shall give Goethe's demands a respectful hearing and confront them with our present-day world, precisely because we do not consider there is so much ground for pessimism. In the century and a half that has passed since Goethe in Weimar brooded and wrote on the ground phenomenon of the origin of colors, the world has developed very differently from what he expected. Nevertheless, as we are bound to maintain against the all too severe critics of our own age, it has not yet been finally carried off by the Devil with whom Faust concluded his fateful bargain. So let us take one more look at the old controversy, with the eyes of today.

For Goethe, all observation and understanding of nature began with the immediate sensory impression; not, therefore, with an isolated phenomenon, filtered out with instruments and so to speak wrung from nature, but with the free natural happening, directly accessible to our senses. Let us take a passage at random from the section on "Physiological Colours" (VI, § 75) in Goethe's *Theory of Colours* (tr. C. L. Eastlake, London 1840; repr., M.I.T. Press, Cambridge, Mass., 1970). The descent from the snowy Brocken on a winter's evening gives rise to the following observation: "During the day, owing to the yellowish hue of the snow, shadows tending to violet had already been observable; these might now be pronounced to be decidedly blue, as the illumined parts exhibited a yellow deepening to orange. But as the sun at last was about to set, and its rays, greatly mitigated by the thicker vapors, began to diffuse a most beautiful red colour over the whole scene around me, the shadow colour changed to a green, in lightness to be compared to a sea-green, in beauty to the green of the emerald. The appearance became more and more vivid: one might have imagined oneself in a fairy world, for every object had clothed itself in the two vivid and so beautifully harmonising colours, till at last, as the sun went down, the magnificent spectacle was lost in a grey twilight, and by degrees in a clear moon-and-starlight night."

But Goethe did not simply stop at immediate observation. He knew very well that it is only by the guidance of a connection, at first merely suspected but then becoming certain with success, that the immediate impression can also become knowledge. Let me cite, for example, a passage from the preface to the *Theory of Colours* (p. xl) : "Surely the mere inspection of a subject can profit us but little. Every act of seeing leads to consideration, consideration to reflection, reflection to combination, and thus it may be said that in every attentive look on nature we already theorise. But in order to guard against the possible abuse of this abstract view [we are afraid of], in order that the practical deductions we look to should be really [vivid and] useful, we should theorise without forgetting that we are so doing, we should theorise with mental self-possession, [with freedom] and, to use a bold word, with irony."

"This abstract view we are afraid of." At this point we already have a precise indication of where Goethe's road must branch off from that of accepted science. Goethe knew that all knowledge has need of images, of combination, of structures that give meaning. Without them, knowledge would be impossible. But the road to these structures leads inevitably, at a later stage, into abstraction. Goethe had already encountered this in his investigations into the morphology of plants. In the immensely varied shapes of the plants he observed, especially on his Italian journey, he thought upon closer examination that he could discern with increasing clearness an underlying principle of unity. He spoke of the "essential form, which nature does but continually play with, as it were, and in playing brings forth the manifold life" (letter to Charlotte von Stein, July 10, 1786), and from hence he arrives at the notion of a ground phenomenon, the archetypal plant. "With this model," says Goethe, "and the key to it, it will be possible to go on for ever inventing new plants. . . . [which, even] if they do not actually exist . . . could [do so, and] . . . possess an inner necessity and truth." (*Italian Journey,* tr. Auden and Mayer, Pt. II, pp. 305–306 [letter to Herder, May 17, 1787].) Here Goethe stands on the frontier of the abstraction he was afraid of. He has himself resolved against crossing this frontier. He has also warned and advised the physicists and philosophers to keep within it as well. "But when even such a primordial phenomenon is arrived at, the evil still is that we refuse to recognize it as such, that we still aim at something beyond, although it would become us to confess that we are arrived at the limits of experimental knowledge. Let the observer of nature suffer the primordial phenomenon to remain undisturbed in its beauty." (*Theory of Colours,* § 177.) The frontiers of the abstract should therefore not be breached. Where the limit of observation is reached, the road should not be prolonged by replacing observation with abstract thought. Goethe was persuaded that detachment from the sensory real world, the entry into this boundless region of abstraction, could lead only to more harm than good.

But since Newton's time, science had already taken other paths. From the first, it has not been fearful of abstraction, and its successes in explaining the planetary system, in the practical use of mechanics, in the construction of optical instruments and in much else, have seemingly justified it, and have resulted in Goethe's warnings going unheard. To this very day, indeed, natural science has developed in a perfectly linear and consecutive manner from Newton's great work, the *Philosophiae Naturalis Principia Mathematica.* Its consequences in technology have transformed the face of the earth.

In this customary type of science, abstraction is employed at two rather different places. The problem, of course, is to recognize the simple in the colorful variety of phenomena. The efforts of the physicist have had to be directed toward sifting out simple processes from the bewildering complexity of appearances. But what is simple? Since Galileo and Newton, the answer has been: A process is simple whose regular occurrence can be represented quantitatively, in all its details, in a mathematically consistent manner. The simple process is not, therefore, that which nature immediately presents to us; on the contrary, the physicist, often by the use of extremely complex apparatus, must first divide up the colorful medley of phenomena, freeing what is important from all unneeded clutter, until the one "simple" process emerges clearly on its own, precisely in order that he may disregard—that is, abstract from—all accessory phenomena. That is the first form of abstraction, and Goethe's view of it is that by this process we have indeed already banished nature itself. He says: "But we greet the bold assertion that this is indeed still nature, at least with a quiet smile, a gentle shake of the head; after all, it does not occur to the architect to pass off his palaces as mountain sides and forests."

The other form of abstraction consists in the use of mathematics in order to represent the phenomena. Newtonian mechanics showed for the first time—and this was the reason for its enormous success—that in mathematical description immense tracts of experience can be gathered into unity and thereby be simply understood. Galileo's laws of falling bodies, the moon's motion about the earth, the revolutions of the planets about the sun, the oscillations of a pendulum, the trajectory of a thrown stone—all these phenomena could be mathematically deduced from the one basic postulate of Newtonian mechanics, the equation "Mass × Acceleration = Force," together with the law of gravitation. The symbolizing mathematical equation was thus the abstract key to a unitary understanding of extensive tracts of nature; and against the confidence in the efficacy of this key, Goethe battled in vain. In a letter to Karl Friedrich Zelter he says: "And it is, in fact, the greatest evil of the more modern physics, that experiments are, as it were, separated from man himself, and that Nature is recognized only in that which artificial instruments demonstrate—nay, they want to prove and limit her capability by these. It is precisely the same with calculation. There is much that is true, that will not admit of being computed, just as there is a great deal that cannot be brought to the test of definite experiment." (*Goethe's Letters to Zelter,* ed. and tr. A. D. Coleridge, 1887; No. 47, p. 62.)

Did Goethe really fail to recognize the ordering power, the cognitive achievement, of scientific method, experiment and mathematics? Did he underestimate the opponent he fought against so tirelessly in the *Theory of Colours* and in many other places? Or was he unwilling to acknowledge this power because values were at stake that he was not ready to sacrifice? We shall have to answer that Goethe was unwilling to tread this abstract path to unitary understanding because it seemed to him too perilous.

Nowhere does Goethe specify the dangers he was afraid of here. But Faust, his most celebrated literary creation, allows us to guess what they were. Faust, among much else, is also a disillusioned physicist. He has surrounded himself in his study with scientific apparatus. But he says:

Ye instruments, forsooth, but jeer at me,
your wheels and cogs mere things of wonder;
when at the door, you were my keys to be,
yet, deftly wrought, your bits can move no wards asunder.

(*Faust,* Pt. I, lines 668–671
[B. Taylor translation])

The mysterious signs Faust seeks in the book of Nostradamus are perhaps the symbols of mathematics, applied in some fashion. And this whole world of symbols and instruments, this insatiable urge for ever wider, deeper and more abstract knowledge, is what causes him, doubter that he is, to conclude his pact with the Devil. Thus the road leading out of natural life into abstract knowledge can end up with the Devil. That was the danger which determined Goethe's attitude to the world of science and technology. He sensed the demonic forces that become active in this development, and felt he should avoid them. But we shall perhaps be compelled to reply that the Devil cannot be escaped so easily.

Goethe himself was soon obliged to make compromises. The most important, no doubt, was his acceptance of the Copernican system, whose persuasiveness even he was unable to resist. But here too, Goethe realized how much must be sacrificed in doing so. I quote again from the *Theory of Colours:* "But among all discoveries and convictions, nothing can have produced a greater effect on the human mind than the doctrine of Copernicus. Hardly had the world been acknowledged as round and closed in upon itself, than it had to renounce the monstrous presumption of being the centre of the universe. Perhaps a greater demand has never been laid upon mankind; for by this admission, how much else did not collapse in dust and smoke: a second paradise, a world of innocence, poetry and piety, the witness of the senses, the convictions of a poetic and religious faith; no wonder that men had no stomach for all this, that they ranged themselves in every way against such a doctrine, which summoned and entitled those who accepted it to a freedom of thought and largeness of mind they had neither known nor even dreamt of before." (Pt. II [Historical], iv, 8.)

This passage will also have to be quoted against all those who, to avoid the dangers feared by Goethe, attempt even now to cast doubt on the correctness and authority of modern science. It is pointed out that this science also changes or modifies its views in the course of time, that Newtonian mechanics, for example, is no longer accepted as correct and has been replaced by relativity and the quantum theory, and that we therefore have every reason to be skeptical of the claims of such a science. But this objection rests on a misunderstanding, as can be seen directly, for example, from the question about the earth's position in the planetary system. It is true enough that Einstein's theory of relativity leaves open the possibility of considering the earth to be stationary and the sun as moving around it. But nothing whatever is thereby altered in regard to the crucial claim of the Newtonian theory, that the sun, with its powerful gravitational field, determines the paths of the planets, and that the planetary system can therefore be really understood only by starting from the sun as midpoint or center of the gravitational forces. It must be particularly stressed at this point that there is certainly no avoiding the conclusions of modern science, once its methods are accepted; and its methods consist in observation, purified into experiment, and rational analysis, which takes on its precise shape in mathematical presentation. If experiment and rational analysis are admitted, the correctness of the results cannot seriously be cast in doubt. But perhaps one can confront them with this question: Is the knowlege so obtained of any value?

If we first try to answer this question, not as Goethe understood it but in accordance with the spirit of our times, admitting with little scruple the argument from utility, we can point here to the achievements of modern science and technology; to the effective removal of all sorts of deficiencies, to the alleviation of the pains of sickness by modern medicine, to the ease of communications, and much else. Assuredly Goethe, who wished to take an active part in life, would have greeted such arguments with much understanding. It is in starting from man's station in this world, from the difficulties that oppress him and the demands that others impose on him, that we shall esteem most highly the possibility of being practically and effectively active in this regard, of being able to help others and in general to improve the conditions of life. We only have to read in Goethe large parts of *Wilhelm Meisters Wanderjahre* or the last sections of *Faust* in order to recognize how seriously the poet took precisely this aspect of our problem. Of the various facets of the world of science and technology, it was the pragmatic that was certainly the most intelligible to him. But here, too, Goethe was not able to rid himself of the fear that the Devil might hereabouts have a hand in the game. In the last act of *Faust,* success, the richness of the active life, is turned into absurdity with the murder of Philemon and Baucis.

But even where the Devil's hand is not so immediately obvious, events are threatened by his power. Goethe recognized that the progressive reshaping of the world by the combination of science and technology could not be halted. He gave anxious expression to the fact in *Wilhelm Meisters Wanderjahre* (III, 13; Artemis Ed., VIII, p. 460): "The growing prevalence of machines distresses and worries me. It is rolling up like a storm, slowly, slowly. But it has taken its course, it will come, and it will strike." Goethe knew, therefore, what lay ahead, and he had thought about how this outcome would react upon the behavior of men. In his *Letters to Zelter* we find: "Wealth and rapidity are what the world admires and what everyone strives to attain. Railways, quick mails, steamships, and every possible kind of facility in the way of communication are what the educated world has in view, that it may [outdo and] over-educate itself, and thereby continue in a state of mediocrity. . . . Properly speaking, this is the century for men with heads on their shoulders, for practical men of quick perceptions, who, because they possess a certain adroitness, feel their superiority to the multitude, even though they themselves may not be gifted in the highest degree." (Tr. Coleridge, 1887, No. 183, pp. 246–247.) Or again in the *Wanderjahre:* "Now is the time of specialties. Happy he who understands this, and works for himself and others in that spirit." (I, 4,; Artemis Ed., VIII, p. 43. Carlyle's translation, Everyman Ed. II, p. 183.) Goethe was thus able to foresee a considerable distance along the road, and looked with the greatest foreboding upon what lay ahead.

In the meantime, nearly a century and a half have passed, and we know where this road has led us to today. Jet aircraft, electronic computers, moon rockets, atom bombs—these are some of the latest milestones we have met by the roadside. The world governed by Newtonian science, which Goethe hoped to stay clear of, has thus become our reality, and it is no help at all to us to be reminded that Faust's partner has also had a hand in the game here. We have to put up with it, as we have always had to do. Nor, indeed, are we anywhere near the end of this road. The time is probably not far distant when biology, too, will be wholly caught up in this process of technical development. That the dangers will then multiply, even beyond the threat posed by atomic weapons, has often been pointed out, most sharply, perhaps, in that pitiless caricature of a future society that Aldous Huxley delineated under the title *Brave New World.* The possibility of breeding men for their appointed tasks, of rationalizing all life on earth through the cult of utility and thereby rendering it meaningless, is here carried through to absurdity with gruesome consistency. But there is no need to go so far in order to recognize that utility as such is not a value in itself but merely shifts the question of value one place over, to the further question: is there any value in the purpose to which the knowledge and powers in question are applied and which they are meant to serve?

Modern medicine has largely eradicated the major plagues from the earth. It has saved the lives of many patients and spared countless people from horrible sufferings, but it has also led to a global population explosion that, if it cannot be slowed down in the relatively near future by peaceful measures of organization, is bound to end in frightful catastrophes. Who can tell whether modern medicine is in all cases setting its goals aright?

Modern science is yielding knowledge whose correctness cannot, on the whole, be doubted; and the technology springing from it allows this knowledge to be harnessed also to the realization of wide-ranging goals. But that by no means settles the question whether the progress thus attained is of any value. This question can be decided only in light of the value concepts that men choose to be guided by in setting the goals. But these ideas of value cannot come from science itself; at all events, that is not where they come from meanwhile. Goethe's cardinal objection to the post-Newtonian methodology of science is thus assuredly directed against the divorce, in this methodology, of the concepts of "correctness" and "truth." For him, truth was inseparable from the value concept. The "unum, bonum, verum," the "one, the good and the true," was for him, as it was for the old philosophers, the only possible compass by which mankind could be guided in seeking its course through the centuries. But a science that is merely correct, in which the concepts of correctness and truth have separated, and hence where the divine order no longer determines the direction by itself, is too greatly imperiled, is too far exposed—to recall *Faust* again—to the clutch of the Devil. So Goethe was unwilling to accept it. In a darkened world no longer illuminated by the light of this center, the "unum, bonum, verum," technical advances—as Erich Heller once put it in this connection—are scarcely more than despairing attempts to make Hell a more agreeable place to live in. This must be particularly emphasized against those who think that by spreading the civilization of science and technology even to the uttermost ends of the earth, they can furnish all the essential preconditions for a golden age. One cannot escape the Devil so easily as that.

Before inquiring whether correctness and truth are really so completely separated in modern science as has hitherto seemed to be the case, we must now put the counterquestion: Does Goethe, with his kind of science, his way of looking at nature, really have anything effective to set against the world of science and technology that has arisen in the aftermath of Newton? We know that in spite of the enormous effect Goethe's poetry exercised in the nineteenth century, his thoughts upon science have become known and fruitful only within a relatively small circle. But perhaps they contain a seed that can develop with careful cultivation, now that the nineteenth century's somewhat naïve belief in progress has given way to a more sober view of the matter. Here we shall again have to ask what the really characteristic features of this Goethean view of nature are, wherein his way of looking at nature differed from that of Newton and his successors.

At this point it will be insisted above all else that Goethe's view of nature starts directly from man: he and his immediate experience of nature form the center whence the phenomena range themselves in an intelligible order. Such a formulation is correct enough, and it makes the major difference between Goethe's and Newton's views of nature particularly clear. But it overlooks, for all that, a most essential point, namely Goethe's conviction that man is visibly confronted in nature by the divine order. It was not the individual's experience of nature, however that may have inspired him as a young man, which was important to the older Goethe, but rather the divine order that becomes perceivable in this experience. It is no mere poetic metaphor for him when, in such a poem as "Bequest of the Ancient Persian Faith" (*West-Eastern Divan, XI*), the believer is moved by the sight of the sun rising over the mountains:

> *God upon His throne then to proclaim,*
> *Him, the life-fount's mighty Lord, to name,*
> *Worthily to prize that glorious sight,*
> *And to wander on beneath His light.*
>
> (tr. E. A. Bowring)

What the experience of nature contains here must be matched, in Goethe's opinion, by the scientific method, and hence the search for the ground phenomenon must be conceived as the pursuit of those God-given structures underlying the appearance, which can be not only construed by the understanding but immediately discerned, experienced and felt. "A ground phenomenon," so he explains (letter to C. D. von Buttel, May 1827), "is not to be equated with a fundamental principle, yielding many kinds of consequences, but is to be regarded as a basic appearance, within which the manifold is to be discerned. Seeing, knowing, sensing, believing and whatever all the feelers may be called, whereby man gropes about in the universe, must then genuinely work together, if we wish to fulfil our important, though difficult, vocation." Goethe very clearly feels that the basic structures must be of such a kind that it can no longer be determined whether they belong to what we think of as the objective world or to the human soul, since they form the presupposition for both. He hopes, therefore, that by "seeing, knowing, sensing, believing," they will come to have effect. But then, we are bound to ask, how do we know, and how does Goethe know, that the real and deepest connections can thus become immediately visible, that they thus lie open to view? May it not be that what Goethe feels to be the divine order of natural appearance is revealed to us in full clarity only at the higher stages of abstraction? Cannot modern science at this point provide answers that could have withstood all Goethe's evaluative demands?

Before going on to discuss such difficult questions, we must say another word about Goethe's rejection of romanticism. In his letters, essays and conversation, Goethe often joined issue at length with romanticism, which was certainly the artistic movement of his day. The same charges are continually being leveled: subjectivism, fanaticism, running to extremes, even to infinity, morbid sensibility, a cult of the antique, sickly prostration and finally sycophancy and duplicity. Goethe's aversion to the apparent sickliness in romanticism, his forebodings about its possible miscarriage, were so strong that he was only rarely able to bring himself to see or even to acknowledge its cultural achievements. All art which, like that of the romantics, alienates itself from the world and no longer seeks to proclaim the reality of that world, but only its reflection in the artist's soul, seemed to him just as disturbing as a science that takes as its study not nature in the free state but the isolated phenomenon, separated out by instruments and prepared for the purpose.

Romanticism can indeed be regarded, in part at least, as the reaction to a world beginning to tranform itself by rationalism, science and technology into a matter-of-fact, practical precondition for outward life, so that it no longer offered any proper scope for the personality in its wholeness, its wishes and hopes and woes. This personality therefore drew back within itself; and the dissolution of the immediate real world, where our acts have consequences that we must face, was indeed felt as a loss; but, so Goethe feared, that also made it easier, not to say more convenient, to take flight instead into a world of dreams, to abandon oneself to the ecstasy of passion, to cast off responsibility for oneself and others and to luxuriate in the boundless expanse of feeling. This step, from an art that seeks to shape the world in its immediate reality to an artistic delineation and exaggeration of the abysses in the human soul, Goethe was no more able to approve than the step into abstraction that science had found itself constrained to take.

In both cases, the affinity between the motives for Goethe's aversion may well go further still. If Goethe was fearful of abstraction in science, if he recoiled in horror at its boundlessness, this was because he thought he detected in it demonic forces to whose menace he did not wish to be exposed. He personified them in the figure of Mephistopheles. In romanticism he felt forces of a similar kind to be at work—again the boundlessness, the loosing of the real world from its wholesome fixed standards, the danger of degeneration into morbidity. It may further have contributed to Goethe's attitude that the highest art form of this next stage was at all times relatively alien to him. Mathematics, which may here be described as the art form of abstraction, was never able to engage or fascinate Goethe, although he respected it. Certainly music, which in German romanticism seems to me to have produced the highest artistic achievements, never moved Goethe as did poetry or painting. We do not know what he would have thought of romanticism if the language of, say, Schubert's C major quintet had been able really to reach him. But he would surely have had to feel that the forces he feared, which operate more strongly in this music than in almost any other romantic work of art, no longer come from Mephisto, no longer proclaim his power but rather the dominion of that bright region whence Lucifer sprang, though it disavowed him. So it is not so remarkable, after all, that here too, in assessing the value of romanticism, the succeeding age did not follow the advice of Germany's greatest poet, or that art turned in large measure instead toward the aims and objects to which romanticism had been the first to dedicate itself. The history of music, painting and literature in the nineteenth century shows how fruitful the romantic approaches have been. To be sure, it also shows, especially if we follow it on into our own century, how justified were Goethe's qualms and objections, just as they have been in the case of science and technology. One may well regard certain oft-bemoaned symptoms of dissolution in the field of art—no less than the use of atomic weapons in technology—as resulting from the loss of that center which Goethe strove all his life to preserve.

But let us return to the question whether the knowledge Goethe sought in his natural science, the knowledge, that is, of the ultimate shaping forces in nature that he felt to be divine, has so completely disappeared from what at first seems the mere "correctness" of modern science.

> *'Tis to detect the inmost force*
> *which binds the world and guides its course,*
> *all germs and forces to explore—*
> *and bandy empty words no more!*
>
> (*Faust,* I, lines 382–385)

That was how the demand ran. On the way thither, Goethe had arrived in his contemplation of nature at the ground phenomenon, and in his morphology of plants at the archetypal plant. This ground phenomenon was not to be a fundamental principle, from which the various phenomena had to be deduced, but a basic appearance within which the manifold was to be discerned. Nevertheless Schiller, at that first famous meeting in Jena in 1794 which began his friendship with Goethe, made it clear to the poet that his ground phenomenon was really not an appearance but an Idea. An Idea, it may be added, in Plato's sense; and since the word "Idea" has acquired a rather too subjective taint, perhaps we now would prefer to read "structure" at this point. The archetypal plant is the primordial form, the basic structure, the shaping principle of plants, which need not only be constructed by the understanding but can also come directly to our awareness in intuition.

The distinction to which Goethe attaches such importance here, between immediate intuition and merely rational deduction, corresponds pretty closely, no doubt, to the distinction between two types of knowledge, *episteme* and *dianoia,* in the philosophy of Plato. *Episteme* is precisely that immediate awareness at which one can halt and behind which there is no need to seek anything further. *Dianoia* is the ability to analyze in detail, the result of logical deduction. It is also apparent in Plato that only *episteme,* the first kind of knowledge, furnishes a connection with the true, the essentially real, with the world of values, whereas *dianoia* yields knowledge, indeed, but knowledge merely devoid of values.

What Schiller was trying to explain to Goethe, on the way home from the scientific lecture they had both been attending, was actually not Platonist but Kantian philosophy. Here the word "Idea" has a rather different and more subjective meaning; and in any case it is sharply distinguished from the appearance itself, so that Schiller's contention that the archetypal plant was an Idea was profoundly upsetting to Goethe. He answered, "I am delighted to find I have Ideas without knowing it, and am able to contemplate them with my own eyes." In the discussion that followed, which, as Goethe tells us, was vigorous in the extreme, Schiller replied, "How can any experience ever be given which should be appropriate to an Idea? For that is precisely what is characteristic of the latter, that an experience could never be adequate to it." (*Biographische Einzelnheiten;* Artemis Ed., XII, pp. 622–23.) From the point of view of Platonism, however, what was at issue in this discussion was really not so much what an Idea is but rather by what organ of knowledge the Idea is revealed to us. If Goethe could see the Ideas with his own eyes, they were certainly different eyes from those that we commonly talk about nowadays. At all events they could not be replaced, in this context, by a microscope or a photographic plate. But whatever decision we reach on this issue, the archetypal plant is thus an Idea, and proves itself to be such, in that by means of it, and using this basic structure as a key, we can, as Goethe says, invent plants *ad infinitum*. By means of it we have, then, understood the make-up of plants; and to "understand" means to trace back to a simple, unitary principle.

But how does all this look in modern biology? Here, too, there is a basic structure which determines not only the shape of all plants but that of all organisms whatsoever. It is a tiny invisible object, a chain molecule, namely the celebrated double helix of nucleic acid, whose structure was elucidated some fifteen years ago by Francis Crick and James D. Watson in Cambridge, and which carries the entire genetic code of the organism in question. In virtue of a great many findings in modern biology, it is no longer possible to doubt that the structure of the living organism is determined by this chain molecule, and that from it, as it were, there proceeds the entire shaping force which fixes the organism's make-up. Naturally, we cannot go into details here. Concerning the correctness of this statement, the same applies as what was said earlier concerning the correctness of scientific statements generally. It rests upon scientific method, upon observation and rational analysis. Once the initial stages of uncertainty in a particular scientific development have been overcome, correctness depends upon the coordinated effect of an extraordinary number of individual facts, upon an immense and complex web of experiences, which gives the statement its unassailable certainty.

Now can the basic structure just described, the double helix of nucleic acid, be in any way compared to Goethe's archetypal plant? The invisible minuteness of this object seems at first to rule out any such comparison. Yet it will be difficult to dispute that this molecule fulfills the same function in the framework of biology as Goethe's archetypal plant was meant to do in botany. For in both cases it is a matter of understanding the shaping, form-giving forces in animate nature, of tracing them back to something simple, common to all living forms. That is just what is done by the ultimate pattern unit in contemporary molecular biology, though it is still rather too primitive to be described as a primal organism. It by no means yet possesses all the functions of a complete living organism; but that need not hinder us from describing it in that or some similar fashion.

This primal unit also has this in common with Goethe's archetypal plant; it is not only a basic structure, an Idea, a notion, a form-giving force, but is also an object or appearance, even if it cannot be seen with our normal eyes but is revealed only indirectly. It can be detected with microscopes of high resolving power and by means of rational analysis, and is thus perfectly real and by no means merely a thing of thought. To that extent it satisfies virtually all the requirements that Goethe imposed on the ground phenomenon—though whether we can "see, feel and sense" it in Goethe's fashion, whether, in other words, it can become an object of *episteme,* of pure knowledge in Plato's sense, may well appear doubtful. Normally, at all events, the primal biological unit is not seen in that way. It is possible merely to imagine that this is how it first appeared to its discoverers.

If we ask, then, about the relationship between correctness and truth in modern science, we shall certainly have to affirm that on their pragmatic side the two concepts are totally separate. But where, as in biology, it is a matter of discerning very broad connections that have been present in nature from the beginning and have not been made by man, we shall be able to establish a certain closing of the gap. For the very broad connections become apparent in the basic structures, in the Platonic Ideas that thereby manifest themselves; and since these Ideas give tidings of the underlying total order, they may also, perhaps, be picked up by other areas of the human psyche than merely that of *ratio*—areas which themselves in turn stand in immediate relation to that total order, and hence also to the world of values.

This becomes especially clear when we pass over to those very general regularities which span the territories of biology, chemistry and physics, and have first become discernible in recent years in connection with the physics of elementary particles. Here we are dealing with basic structures of nature or of the world at large, which lie still deeper than those of biology and are therefore still more abstract, still less directly accessible to our senses than the latter. But they are also to that extent more simple, for they have only to represent the universal and no longer the particular at all. Whereas the primal structure of biology has not only to represent the living organism as such but must also distinguish the numberless different organisms—by means of the various possible arrangements of a few chemical groups along the chain—the basic structures of all nature have only to represent the very existence of nature itself. In modern physics, this conception is realized in the following manner: We formulate in mathematical language a fundamental law of nature, a world formula, as it has sometimes been called, which all phenomena in nature must satisfy, and which therefore symbolizes in a fashion the mere possibility, the existence, of nature. The simplest solutions of this mathematical equation represent the various elementary particles, which are basic forms of nature in exactly the same sense as that in which Plato considered the regular solids of mathematics—cubes, tetrahedra, etc.—to be such basic natural forms. Again, to recur to the debate between Schiller and Goethe, they are also "Ideas," just as Goethe's archetypal plant was, even though they cannot be seen with normal eyes. Whether they can be intuited in Goethe's sense is doubtless simply a matter of the cognitive organs we use to approach nature. That these basic structures are intimately connected with the order of the macrocosm as a whole is surely almost beyond dispute. But it remains up to us whether we wish merely to seize upon the one narrow, rationally apprehensible segment from this immense system of connection.

Let us cast one more backward glance at the course of historical development. In science as in art, the world since Goethe's day has gone the way he warned us against, since he considered it too dangerous. Art has withdrawn from the immediately real into the interior of the human soul, while science has taken the step into abstraction, has conquered the huge expanse of modern technology and has pushed on to the primal structures of biology and the ground forms that correspond in modern science to the Platonic solids. At the same time, the dangers have become fully as threatening as Goethe foresaw. We have in mind, for example, the soulless depersonalizing of labor, the absurdity of modern armaments, the flight into insanity that took the form of a political movement. The Devil is a powerful fellow. But the lucid region we spoke of earlier in connection with romantic music, and which Goethe was able to discern throughout all nature, has also become visible in modern science, at the point where it yields intimations of the mighty unity in the ordering of the world. Even today we can still learn from Goethe that we should not let everything else atrophy in favor of the one organ of rational analysis; that it is a matter, rather, of seizing upon reality with all the organs that are given to us, and trusting that this reality will then also reflect the essence of things, the "one, the good and the true." Let us hope that the future will be more successful in this regard than our own day, than my own generation, has found it possible to be.

XI

The Tendency to Abstraction in Modern Art and Science

The general theme of our symposium today is as follows: the significance of modern scientific knowledge—in medicine, physiology and physics—for art and art education, especially music and musical education. I have no wish to discuss this subject here from the technical point of view; it would naturally be possible for a physicist to start out from modern acoustics, say from the production of sound in electronic instruments, and thence to draw conclusions in regard to modern music. Instead, I should like to consider the subject rather more in principle, or, if you will, in terms of the philosophy of culture, and to ask whether—as is often maintained—the tendencies in modern art, and particularly modern music, which often appear so strange and unintelligible, can be shown to have a parallel in the form of similar phenomena in modern science; and whether by a comparison with modern science we can learn something about these peculiar phenomena. We shall not be concerned, therefore, with particular forms or techniques of modern art or science but with their overall shape.

It is frequently said that contemporary art is more abstract than the art of the past, that it is further removed from actual life, and that this is what links it to modern science and technology, of which the same is also true to a very high degree. I

Address delivered at a symposium of the Karajan Foundation in Salzburg, 1969.

should like to leave aside for the moment the question of how far the term "abstract" may be a correct description of these phenomena in contemporary art. What is certain is that in modern science, abstraction plays an altogether decisive part. I should like, therefore, to start with a brief description of the process, to make clear its inherent inevitability and to show that it has been crucial to the tremendous advances made by modern science, so that nobody in any way interested in scientific progress could wish to see the process put into reverse. I shall then try to discover whether anything comparable is occurring, or has occurred, in the development of modern art. I shall confine myself here to pursuing the oft-cited comparison rather more exactly than is commonly done. In so doing I must emphasize that I am not really competent for the task; I know the history of art only at second hand, have not studied it in detail and am therefore in danger of judging superficially. I am also well aware that in this comparison I can deal only with a very small and perhaps unimportant section of the broad range of problems included in the term "abstract art." But the purpose is only to provide incentives to discussion.

As my first example of the tendency to abstraction in modern science, I should like to instance the development of biology. In earlier times, say at the end of the eighteenth century, biology consisted of the two subordinate fields of zoology and botany; scholars described the manifold forms of organisms, established similarities and differences, constructed affinities and endeavored to bring system into the abundance of phenomena. But even this science, dedicated to life as it was immediately encountered, could not resist the search for unitary viewpoints from which to understand collectively the varied forms of life. Thus even Goethe, who, as he said, was fearful of all abstraction, went looking for the *Urpflanze,* the prototype of plants, from which all others could be derived and understood. Schiller had to go to great trouble to explain to Goethe that the *Urpflanze* is an Idea, the Idea of a plant, and thus to that extent already an abstraction. The succeeding period began by perceiving the element common to the various organisms in the different biological functions, such as metabolism and propagation. It then inquired into the physicochemical processes whereby these functions are realized in the organism, and was thus inevitably pushed on toward the smallest parts of the organism, and so to molecular biology. The ground structure common to all organisms was finally recognized in recent years to be a chain molecule, a nucleic acid, which can be observed under microscopes of maximum resolving power and shown to be a basic component of all living matter. On this chain molecule, whose details are of no concern to us here, the whole inheritance of the organism in question is written in a chemical script, and according to this code the new organisms are formed in the process of propagation. We can, if we like, compare this chain molecule, the nucleic acid, with Goethe's *Urpflanze;* but even so, molecular biology still represents a concern with chemical structural formulae of the most complex kind, to which we cannot have the immediate relation that we have to the individual organisms.

From the historical development here briefly sketched, it is already possible to see clearly the elements responsible for the tendency to abstraction. To understand means to recognize connections, to see the individual as a special case of something more general. But the step toward greater generality is always itself a step into abstraction—or more precisely, into the next highest level of abstraction; for the more general unites the wealth of diverse individual things or processes under a unitary point of view, which means at the same time that it disregards other features considered to be unimportant. In other words, it abstracts from them.

The same process has occurred in other sciences, in chemistry and physics; but from this development I should like to single out only two particular episodes from the evolution of modern atomic physics, because later I shall be able to bring them into comparison with what may perhaps be a corresponding development in modern art. Twice in the present century, enormous expansions of physics have been effected; in the revision of the structure of space and time by the theory of relativity, and in the formulation of authoritative laws for atomic physics by the quantum theory. In both cases the newly emergent physics has been very much less intuitive, and in that sense more abstract, than what went before. A genuine step into a higher stage of abstraction has been taken. In both cases, moreover, the move to a higher level of abstraction was preceded by an extraordinary intervening period of uncertainty and confusion that lasted for several years. I should like to describe this intervening stage in rather more detail.

In relativity theory, the uncertainty began when the attempt was made to establish the earth's motion in space by electromagnetic means. The concept of motion became unclear. Did we mean the earth's motion relative to the sun, or relative to other star systems or relative to space? And is there any such thing as motion relative to space? Then the concept of simultaneity became unclear. The question was raised whether we know what is meant in saying, for example, that an event in the Andromeda nebula is simultaneous with one on earth. It was felt that we no longer knew this; but nobody was yet in a position to formulate the real connections precisely. Even worse was the period of uncertainty and confusion prior to the emergence of the quantum theory. We could follow the tracks of electrons in a cloud chamber; so obviously there were electrons and electron tracks; but in the atom there seemed to be no such tracks. It was felt that we no longer knew exactly what was meant by the words "position" or "velocity" of an electron within the atom; but for a long time we were not able to speak of processes within the atom in any rationally intelligible fashion. In the development of quantum theory, the period of uncertainty and confusion lasted for a quarter of a century, and was thus by no means a rapidly surmounted transitional phase.

Perhaps a word should also be said about the men who in those days were working in this field. They were in despair about the state of confusion and aware that no lasting knowledge could come of it. Nobody wanted to destroy or repudiate the older physics. But the physicists were set a task which could no longer be evaded; for eventually it would have to be possible to state, in precise, rational language, what was happening in the interior of the atom. But this could obviously no longer be done in terms of the older physics. There was thus a content that had to be shaped, namely the results of numerous experiments on atoms. The interconnection manifestly present among these many experiments would have to become clearly expressible. But for a long time this goal was too difficult. It was not until success had been finally achieved that physicists felt their subject was in order again. For those who took part in it, the emergence of the new order was a powerful and astonishing experience. The physicists working in the critical area felt at once that something most excitingly new and wholly unexpected was taking place. But I do not want to dwell any longer on the description of such experiences. I should like only to mention one further point, which is important for the comparison to come. In all these matters, mathematics, and especially its modern technological form, its execution by electronic computers, played only a subordinate, secondary role. Mathematics is the form in which we express our understanding of nature; but it is not the content of that understanding. Modern science and, I think, also the modern development of art are misunderstood at a crucial point if we overestimate the significance in them of the formal element.

I should now like to move over into a comparison of the picture thus briefly sketched with what happens in the development of art. To begin with, I take up the problem just mentioned, concerning form and content, which in my opinion occupies a central position here. Art has a different task, of course, from that of science. Whereas science explains and makes intelligible, art has to present, to illuminate, to make visible the basis of human life. But the problem of form and content is similarly posed in both areas. The progress of art consists, no doubt, in the manner whereby initially a slow historical process, transforming men's lives without the individual's being able to exert much influence upon it, brings forth new contents. Such contents in antiquity, for example, will have been the brilliance of gods conceived as heroes; in the late Middle Ages, the religious security of men; toward the end of the eighteenth century, the world of feeling that we know from Rousseau and Goethe's *Werther*. Gifted individual artists then try to give these contents visible or audible shape by wresting new possibilities of expression from the material, the colors or instruments their art employs. This interplay, or struggle, if you will, between the content to be expressed and the restricted means for expressing it seems to me to be—much as it is in science—the unavoidable precondition for the emergence of genuine art. If no content presses for expression, the soil on which art may grow is lacking; if there are no limitations on the means of expression—if in music, for example, any desired sound can be produced—this struggle no longer exists, and the artist's effort is to some extent a beating of the air.

Now, knowing this, how are we to judge the development of modern art and to compare it with the development of modern science? We at once encounter profound differences here. It is well known that particular trends in modern art are defined by the negation of determinate forms; there is talk of "atonal" music or "nonobjective" painting. Here there is no reference to content, and a reference to form only by way of negation. There has hardly been any comparable process in modern science. We have perhaps talked at times of nonclassical physics, but nobody would have referred in that way to a genuine scientific discipline. The absence of specific forms can never, I think, truly characterize either an art or a science; for it is surely of the nature of such endeavors of the mind that they give shape to a content, and therefore create forms. Of course, there are also large areas of modern art that are otherwise described than by the negation of forms.

A further difference, too, becomes clear here. In modern science the questions have always been set for us by the historical process, with the efforts of the scientist having been directed to answering these questions; whereas in modern art the putting of the questions itself seems unclear. Or, as we might also phrase it: In science there is never any question about *what* requires explaining, but only about *how* it should be explained; in art, on the other hand, it seems to be a problem nowadays as to *what* should be represented—there are too many rather than too few answers on how it should be done. Thus, in contrast to modern science, it often seems in modern art as if the very content that is to be presented is still contested or cannot be grasped. If we wish to answer the question why everything goes on so differently in modern art from the way it does in science, we must therefore pose the question about content in all its sharpness. What could or should be the content of this art of today?

Art, to be sure, has at all times depicted the spirit, the life basis or vital feeling, of the epoch in question. Hence we must inquire what the vital feeling of the contemporary world may be, and especially of its younger generation. Here we observe at once a striving for enlargement, for expansion, which in this form was alien to earlier periods. The young man no longer sees his life merely in relation to the tradition, the country, the culture in which he has grown up, but relates it to the whole world, which at bottom he may think of as a unity. The tendency to feel the earth or the universe to be the habitat to which our own destiny is related will certainly become stronger still in the future. It answers to the tendency in science to regard the whole of nature as a unity and to formulate laws that hold good throughout all its spheres. As was explained earlier, the realization of this program has pushed the sciences on to ever higher levels of abstraction, and to that extent it might well be imagined that the relation of our life to the whole spiritual and social structure of the earth will also be capable of artistic presentation only if we are ready to enter into regions more remote from life.

But besides this tendency to an enlargement of the habitat for the individual, a more negative feature, described in detail by psychologists, is becoming apparent in the life sense of the younger generation. We may call it a striving away from shape, an urge toward "unshaping." This trait becomes plainly discernible, for example, in jazz music and its successors, which some of our young people are very fond of and which is often felt almost as a sort of world outlook. What is characteristic here is the obliteration of contours, both harmonic and rhythmic; the tone is no longer required to be pure, it has to be blurred; the rhythm is divided into a bass rhythm and a melody rhythm, and thereby shifted out of the balance customary in earlier music. In the singing, which instead of a vocal lyric eventually gives utterance merely to disconnected syllables or evocative noises, the form of language is likewise in dissolution. No new form is being advanced here. These features of jazz, the psychologists tell us, are typical of the state of mind of the young. All their feelings display a strikingly blurred, hazy and indistinct quality; and this lack of clarity rests upon a loss of personal and tangible contacts, i.e., an alienation from reality, and at the same time promotes and increases this alienation. As Gunther Anders has said of our young people: "Till a late hour, they do not fit neatly into this world."

I want to suggest that this aspect of our present-day sense of life, or more accurately of the vital feeling of our young people, is a central problem of modern art. At first it seems inevitable that this tendency to unshaping must be the negation of all art, that from this point every road to art is radically obstructed; seeing that art is shaping. The tendency to unshaping cannot be remedied by any experimenting with new forms, computerized methods of composition and the like; for where no content is calling to be shaped, it does not help to invent new forms.

And yet perhaps one can also find parallels to this process of "unshaping" in the growth of modern science, and especially of atomic physics. I have mentioned already that, prior to the formulation of relativity theory and the quantum theory, an extraordinary period of confusion ensued, in which physicists felt that all the concepts with which they had otherwise been at home in the field of nature would no longer fit properly, and that one could now employ these concepts only in a rough and hazy fashion. Naturally, this stage was not one of satisfactory science; it was rather the negation of science. Everyone knew that it could not produce any lasting results; but as preparation for the shaping that was to follow, it had a crucial function: it created the free space that was needed in order to approach the abstract concepts whereby it was later possible to order the wide connecting areas of the field.

It may also be the case today that the tendency to unshaping springs from a sense of life that not only seems to perceive the unreliability of all past forms but also descries behind the forms connections that later, perhaps, may again be able to support life. This may possibly be the most important content of modern art.

Once granted this analogy between the phases of development in modern art and modern science, we recognize that strictly we should not yet be talking nowadays of abstract art. Genuinely abstract art has existed earlier—as in the arabesque ornamentation of the early Middle Ages, or in Bach's *Art of Fugue*—and it will doubtless exist again. Major portions of modern art would be better described, however, as a blurred, indeterminate art, or, as it frequently labels itself, an art of denial and dissolution, though it should be noted here that even the statements of such an art continue to depend upon the old forms, which still glimmer through obscurely at this point. Pure chaos is totally without interest.

If we assess the tendencies of modern art from the standpoint of this analogy—and I have already stressed that we proceed, in so doing, from a very special point of view—the tendency toward universality would have to be reckoned the most strongly positive feature. Art can no longer bind itself to the tradition of any particular culture, but seeks to present a sense of life which perceives man in relation to the whole earth, and sees earth in the cosmos as if from other stars. That a trend of this sort can still also make use of the traditional formal means is evidenced in the writing of the French airman Saint-Exupéry. Form is never of importance in comparison with content. A new language, presenting this universal life sense immediately and bindingly for everyone, has yet to be discovered, perhaps because it is still not really intelligible; but from the best works of contemporary art we can see in what direction this language must be sought.

In all probability, much of what is now occurring in art, indeed the very things that alienate us there, can be compared with that confused prior stage familiar to us in science, which, however disturbing in its details, is what creates space for a knowledge of the new connections and the new language. To that extent, then, one may feel quite optimistic for the future of art, since these prior stages come to an end and lead on into a period of clear shaping. The things that have happened should not, however, be overinflated by grandiose phrases. We read, for example, in a major work on twentieth-century painting: "We have, indeed, destroyed with our life plan the cultures of the world; but the dead still survives and works in the tissues of the living." These strike me as words too grand for a bad business; and the same business was also a bad one for science. It needs the greatest efforts to find the way back into order from here.

The topic of our present symposium is "The Significance for Art of the Findings and Discoveries of Modern Science." So in seriously considering the analogy we have many times alluded to, perhaps a few further points should be stated which we scientists have noted in the development of our subject and which might be of value for the future development of art.

I shall begin these concluding remarks with a question relating to a vogue word of our time, the term "revolution." So often we hear talk of revolutions in science, revolutions in art, revolutions in society. But how does a revolution in science come about? The answer: By trying to change as little as possible; by concentrating all efforts on the solution of a special and obviously still unsolved problem, and proceeding as conservatively as possible in doing so. For only where the novel is forced upon us by the problem itself, where it comes in a sense from outside and not from ourselves, does it later have the power to transform. And then, perhaps, it will bring very extensive changes in its train. Our experiences in science have taught us that nothing is more unfruitful than the maxim that at all costs one must produce something new. Thus in our own science, atomic physics, the seeking out of new formal possibilities, new mathematical schemata, has never produced much unless and until the content of the new connections had become visible. It would be still more unreasonable to suppose that we ought to destroy all the old forms, and that the new will then already emerge of its own accord. With a rule of that kind we should certainly never have gotten anywhere in science: first, without the old forms we could never have found the new; secondly, nothing ever happens, in science or in art, of its own accord—we have to shape the new ourselves. Finally, we should not forget the warning that although ultimately we are concerned with new shaping and the creation of new forms, these can arise only from new content; it can never be the other way round. To create new art, therefore, means, I should suppose, to make new contents visible or audible, not merely to invent new forms.

Perhaps I may once more summarize the substance of this paper in a few sentences, which are intended to stimulate discussion. In art, as in science, we can discern a striving for universality. In the sciences we are endeavoring to interpret all physical phenomena in a unified way, to understand all organisms in terms of a single point of view, and we have already come a long way upon this road. In art we are seeking to present a basis for life common to all men on earth. This striving for unification and bringing together necessarily leads to abstraction, in art probably no less than in science. However, what we now see before us in modern art may well not yet belong to this stage of abstraction. Instead, it probably corresponds to that confused preliminary phase which also had to be undergone in science: a phase where we sense that the previous forms will not suffice to present the new and more embracing content; where this content can be seized but cannot yet be formed, since it is not yet clear or viable enough for that.

Such is the picture for one who judges from the standpoint of the history of science but who also realizes that this judgment may be both superficial and unfair.

XII

Changes of Thought Pattern in the Progress of Science

I shall be dealing in what follows with changes of thought pattern in the progress of science. I must confess that I had originally contemplated a rather more aggressive formulation of my theme. For my topic I wanted to choose "How Does One Make a Revolution?" But I then became apprehensive that you might expect rather too much of my lecture, and perhaps also that I might attract the wrong kind of audience. So I have let it go with the more cautious title "Changes of Thought Pattern." It must be admitted, though, that during the last hundred years there have been such radical changes of thought pattern in the history at least of our own science, physics, that it is perfectly legitimate to speak of one or even several revolutions; and it is in this sense, of "a change in thought pattern," that I shall use the word "revolution" here.

Perhaps I should begin with a historical sketch of the changes in thought pattern that have occurred since the days of Newtonian physics. It is reasonable to take that physics as a starting point, for it was here that the methodology of modern science—experiment and exact description of the phenomena and their connections—was first created and developed. For example, at that time there was a great interest in the motion of bodies under the influence of forces. Thanks to the great success of New-

Lecture delivered to the Association of German Scientists in Munich, 1969.

tonian science and the often (though not always) palpable soundness of its assertions, the idea arose that it would eventually be possible to understand all physical phenomena in terms of this conceptual scheme. The most important concepts were those of time, space, bodies, mass, position, velocity, acceleration and force. Force was an effect of one body upon another.

For a time it proved possible to make significant extensions to Newtonian mechanics, even while adhering to this conceptual scheme. Hydrodynamics, for example, was generated from Newtonian mechanics, merely by giving a rather more general interpretation to the notion of body. Water is naturally not a rigid body. But the individual volumes in the fluid could still be regarded as bodies in the sense of Newtonian physics, and thus it proved possible to find a mathematical representation of the kinematics and dynamics of fluids that was confirmed by experience. People became accustomed to a mode of thought that constantly inquired about the motions of bodies or minute parts of matter under the influence of forces.

Not until the nineteenth century did this type of thought and inquiry encounter its limits. The difficulties arose in two different places and in very different fashion. In the theory of electricity, the concept of a force exerted by one body on another turned out to be inadequate. It was Faraday especially who pointed out that we gain a better understanding of electrical phenomena if we view the force as a function of space and time, if we set it alongside the distribution of velocities or stresses in a fluid or elastic body—in other words, if we go over to the idea of a force field. Such a transition seemed acceptable from the standpoint of Newtonian physics only if there was assumed to be a substance called the ether, equally distributed throughout the cosmos, whose field of stress or torsion could then be identified with the force field of electrodynamics. Without such a hypothetical ether, electrodynamics could not be interpreted from within the Newtonian conceptual framework. Only as the years passed was it realized that this hypothetical ether was really quite unnecessary, that it neither can nor should make any appearance whatever among the phenomena, and that it is therefore more correct to ascribe to the force field a physical reality of its own, independent of all bodies. But with the introduction of such a physical reality, the framework of Newtonian physics was finally burst asunder. Other questions had to be asked than those that could have been put in the earlier physics. Quite generally we may say that a change of thought pattern becomes apparent when words acquire meanings different from those they had before and when new questions are asked.

The second place at which the insufficiency of the old Newtonian scheme of concepts became apparent was in the theory of heat; though here the difficulties were much more subtle and less easy to perceive than in the theory of electricity. At first everything seemed to be going smoothly. Statistics could be applied to the motions of large numbers of molecules, thereby rendering the laws of the phenomenological theory of heat intelligible. Only when there was also an effort to justify the hypothesis of randomness appropriate to such a statistics was it observed that the bounds of Newtonian physics would have to be left behind. Probably the first man to see this with full clarity was J. Willard Gibbs, in the 1870s. But it took decades before the Gibbsian view of heat theory gained acceptance, to some extent at least, and perhaps it still seems strange and unintelligible to many people even today. At all events, its understanding calls for a change of thought pattern, because in it one encounters the concept of an observation situation that is lacking in Newtonian physics, and because one therefore frequently asks other questions without becoming aware of it.

The really radical changes in the foundations of physical thought were first forced upon us, however, only in the twentieth century, by relativity theory and the quantum theory. In relativity theory it turned out that the concept of time in Newtonian mechanics is no longer applicable when dealing with phenomena in which motions of very high velocity play a role. Since the independence of space and time had been among the basic presuppositions of earlier thinking, this pattern of thought had to alter if recognition was to be given to the relations between space and time called for by the theory of relativity. The absolute concept of simultaneity, presupposed as self-evident in Newtonian mechanics, had to be abandoned and replaced by another, depending on the state of motion of the observer. The criticisms of relativity theory so frequently expressed, its obdurate rejection by certain physicists and philosophers, have their origin at this point. These men felt that the change of thought pattern called for here was simply insupportable. But it is nonetheless the prerequisite for an understanding of contemporary physics.

Eventually, much greater demands still were imposed in the quantum theory. The whole objective description of nature in the Newtonian sense, in which determinate values are attributed to the defining elements of the system, such as position, velocity and energy, had to be abandoned in favor of a description of observation situations, in which only the probabilities of certain outcomes can be given. The words in which we allude to atomic phenomena therefore became problematic. It was possible to speak of waves or particles, and necessary to realize at the same time that this expedient by no means involved a dualistic description of the phenomena but rather an absolutely unitary one; the meaning of the old terms became somewhat blurred. It is well known that even such eminent physicists as Einstein, Max von Laue and Erwin Schrödinger were not ready or else not in a position to make this change in the pattern of their thought.

On the whole, then, it is possible to conclude in retrospect that in this century there have been two great revolutions in our science, which have displaced the foundations of physics and thereby altered the whole edifice of the subject. We must now ask how such radical alterations have come about, or—to put it in more sociological terms, though also quite misleadingly—how was a seemingly small group of physicists able to constrain the others to these changes in the structure of science and thought? It goes without saying that the others at first resisted the change, and were bound to do so. I must equally forestall at this point a very natural objection, which is only partially justified here. It might be asserted that this comparison of a revolution in science with a social revolution is completely beside the point, since in science we are ultimately concerned with what is true or false but in society with what is more or less desirable. This objection may be warranted in part. But it will have to be conceded that in the social field the concepts of "true" or "false" could also be replaced by "possible" and "impossible"; for under given external circumstances, not every social form will by any means be possible. Historical possibility is thus an objective criterion of correctness, as experiment is in science. But however that may be, we have to ask how these revolutions came into being.

I shall begin with the history of the quantum theory, since I know it in greatest detail. Once the conviction had been reached, in the last third of the nineteenth century, that both the statistical theory of heat and also electromagnetic radiation had been completely understood, it was inevitably concluded that there was also bound to be success in deducing the law of so-called "blackbody" radiation. But unexpected difficulties arose at this point and gave rise to a feeling of uncertainty. The immediate application to radiation theory of the otherwise demonstrably reliable laws of statistical thermodynamics led to an absurd result that could not possibly be correct. This situation, however, did not have the effect of causing a physicist or group of physicists to sound the alarm and call for the overthrow of physics. There was no talk of such a thing. For the good physicists knew that this edifice of classical physics was so firmly fashioned, so securely anchored together by thousands of experiments, that a violent change could only lead to contradictions. So people did the most reasonable thing that can be done, initially, in such cases—they waited to see whether further developments would not yield new viewpoints that might lead to a solution of the difficulties within the framework of classical physics. Among those who concerned themselves with such problems was one physicist, of a highly conservative turn of mind, who was not content with mere waiting but believed that by an ever more careful, ever more thoroughgoing analysis of the problem it might perhaps be possible to arrive at this new point of view. He was Max Planck. He too had no desire to upset classical physics; he simply wanted to obtain light on this one obviously unsolved problem of "blackbody" radiation. He finally discovered to his horror that in order to interpret this radiation, he was obliged to frame a hypothesis which did not fit into the framework of classical physics and which, from the standpoint of this older physics, seemed completely insane. Later he tried to moderate his quantum hypothesis in order to make the contradictions with classical physics less blatant. But there he had no success.

Only then was the next step taken, which announced the beginning of a real revolution. Einstein showed that the features in Planck's quantum theory which contradicted classical physics were also to be seen in other phenomena, for example in the specific heat of solids or in the emission of light. From there on, the quantum theory broadened out into atomic structure, chemistry and the theory of solids, and in more and more places it was recognized that the quantum hypothesis obviously described an essential feature of nature that had hitherto been overlooked. People were beginning to accept the fact that, for the time being at least, unavoidable inner contradictions were making a real understanding of physics impossible.

You know how matters went after that. Only at the end, in the middle twenties, did it become clear how radical was the reconstruction that had to be undertaken throughout the entire edifice of physics, and especially in its foundations. Only at this time, too, did strong resistances make themselves felt against the completed theory. Until then it had really not been customary to take the quantum theory with complete seriousness, for it was still full of internal contradictions, and so surely could not be final. But from the second half of the twenties onward, it was complete and consistent. Anyone who wanted to understand it was obliged, in the field of physics at least, to alter the pattern of his thinking; he had to put new questions and employ other intuitive pictures than those used earlier. As you know, this created the greatest difficulties for many physicists. Even Einstein, von Laue, Planck and Schrödinger were not prepared to admit the finality of the new situation after the revolution. But I emphasize once more that at no time during this history of the quantum theory was there a physicist or group of physicists seeking to bring about an overthrow of physics.

Now let us compare this evolution of the quantum theory with other, earlier revolutions in the history of physics. Let us ask, then, how did the theory of relativity come into being? The starting point here was the electrodynamics of moving bodies. Since the Hertzian waves were regarded as oscillations of a hypothetical medium, the ether—or were necessarily viewed in terms of the Newtonian conceptual scheme—it had to be asked what went on when, in course of an experiment, there are bodies involved which move relatively to the ether. Here some exceedingly obscure suggestions were put forward that seemed wrong if only because of their complexity. It would be very tempting at this point to speculate on when a proposed formula looks wrong and when not. But I shall refrain from doing so, and would prefer to recall that even at that time the concept of "motion relative to the ether" seemed suspicious to many physicists, since the ether could never otherwise be observed. The physicists had the impression of having somehow strayed into the wilderness, and there was therefore satisfaction that the earth's motion relative to the ether could be investigated by means of the celebrated Michelson-Morley experiment. The result, as we know, was that here too the ether remained undetected, and that a general skepticism subsequently spread among physicists in regard to the notion of an ether and the calculations based upon it.

But again there was no group of physicists who sounded the alarm and proclaimed the downfall of the existing physics. On the contrary, efforts were made to find a solution in terms of the physics of the day, and at all events to make as few changes in it as were humanly possible. Accordingly, H. A. Lorentz proposed to introduce into moving systems an apparent time, correlated with the time measured in a static system by means of the celebrated Lorentz transformation; and to assume that this apparent time was operative for differences of speed between different rays of light. Only then did Einstein observe that the whole picture became infinitely simpler if the apparent time of the Lorentz transformation was identified with the actual time. But thereupon the Lorentz transformation became a statement about the structure of space and time. If this statement is taken to be correct, the words "space" and "time" mean something other than they do in Newtonian physics. The concept of simultaneity was relativized, and the pattern of our physical thinking, which certainly presupposes the concepts of "space" and "time," was altered. To this revolution, likewise, strong resistances subsequently made themselves felt and have instigated innumerable debates about the theory of relativity. At the moment, however, I am concerned only to stress that this revolution in physics also came about without anyone intending to destroy or radically alter the edifice of classical physics.

If we move a step further back into history, to Maxwell's theory and the statistical theory of heat, we are scarcely aware nowadays that even at that time it was a question of profound changes in the pattern of physical thought. By now we are hardly able to view these changes independently of those which took place later in relativity theory and the quantum theory. The introduction of the field concept by Faraday and Maxwell was the first step, as it were, toward the introduction of the field as an independent physical reality by the later creation of the notion of an ether, and in Gibbs' version of the statistical theory of heat there was already an anticipation of the concept of the observation situation, which later played such a decisive role in the quantum theory. That important changes in the pattern of physical thinking were involved here is again to be seen most clearly, perhaps—after the event—in the protracted resistance put up against these theories. But this aspect of the problem will engage us only later on. In both these cases the same holds true as was previously said of the relativity and quantum theories: at no stage of the development did any physicist contemplate the overthrow of the existing order in physics. On the contrary, it was long hoped that the new phenomena could be understood within the framework of Newtonian physics, and only in the final phase did it appear that the fundamentals of the subject had been shifted.

And now a word about the strong resistances that have arisen to every change in the pattern of thought. Anyone working in science is accustomed, as life goes on, to becoming acquainted with new phenomena or new interpretations of phenomena, and even perhaps to making such discoveries himself. He has been put in readiness to fill his mind with new ideas. So he simply cannot wish to hang on conservatively to old accustomed ways. The progress of science therefore continues, as a rule, without much in the way of resistances and disputes. But the case is altered when new groups of phenomena compel changes in the pattern of thought. Here, even the most eminent of physicists find immense difficulties. For the demand for change in the thought pattern may engender the feeling that the ground is to be pulled from under one's feet. A researcher who for years has achieved great success in his science with a pattern of thinking he has been used to from his young days cannot be ready to change this pattern simply on the basis of a few novel experiments. In the most favorable case here, after years of mental wrestling with the new situation, a change of outlook may ensue and open a road into the new way of thinking. I believe that the difficulties at this point can hardly be overestimated. Once one has experienced the desperation with which clever and conciliatory men of science react to the demand for a change in the thought pattern, one can only be amazed that such revolutions in science have actually been possible at all.

How, then, have they come about? The answer that comes

readiest to hand, though it is probably still inadequate, would run as follows: because there is a "right" and a "wrong" in science, and because the new ideas are simply right and the old ones wrong. This answer presupposes that in science it is always the right answer that prevails. But that is by no means the case. For example, the correct notion of a heliocentric planetary system, developed by Aristarchus, was abandoned in favor of Ptolemy's geocentric viewpoint, although the latter was false. Another explanation for the success of revolutions would naturally be more inadequate still: that they come to pass because scientists gladly defer to the authority of a strong revolutionary personality, such as Einstein. There is nothing in that notion, for the internal resistances against a change in the thought pattern are much too strong to be overcome by the authority of any one man.

The correct explanation is surely this: because scientists perceive that with the new pattern of thought they can achieve greater success in their science than with the old; because the new system proves to be fruitful. Once anyone has chosen to be a scientist, he wants above all else to get ahead, to be on hand when new roads are opened up. It does not satisfy him merely to repeat what is old and has often been said before. Hence he will be interested in the kind of problems where, so to speak, there is "something to be done," where he has a prospect of successful work. That is how relativity theory and the quantum theory came to prevail. In the last resort, admittedly, a pragmatic criterion of value is being set up here, and one cannot be absolutely certain that the right will always triumph. Ptolemaic astronomy is again the renowned counterexample. But at all events there are forces at work here that can prove stronger than the internal resistances to a change in the pattern of thought.

Let us return once more from the final stage of a scientific revolution to its outset. From the examples adduced it can be seen, I think, that never in its history has there been a desire for any radical reconstruction of the edifice of physics. At the beginning, rather, we always encounter a very special, narrowly restricted problem, which can find no solution within the traditional framework. The revolution is brought about by researchers who are genuinely trying to solve this special problem but who otherwise wish to change as little as possible in the previously existing science. It is precisely the wish to change things as little as possible which demonstrates that the introduction of novelty is a matter of being compelled by the facts; that the change of thought pattern is enforced by the phenomena, by nature itself, and not by human authorities of any kind.

Is it permissible to carry this analysis over to other revolutions, as in art or in society? I now come back, in conclusion, to the question I posed at the beginning: "How does one make a revolution?" And I shall assume for a moment, experimentally as it were, and without arguing with the historians, that the answer may be valid in all fields at once. It would then run: by trying to change as *little* as possible. If one has recognized that there is a problem incapable of solution within the traditional framework, then it is necessary, so it would seem, to concentrate all efforts on the solution of just this one problem, without at first thinking of changes in other areas. It is then—in science at least—that there is the greatest probability that a true revolution may come of it, so far as there is any necessity at all for new foundations. But that is just what we presupposed, and without this necessity it is quite certain that nothing comparable to a revolution will occur. I may leave it to the historians among you to consider whether the answer I have just given is also valid in history. Still, as an example for this point of view I could cite Martin Luther's reformation of the Church. The need for reform in the Church of that time was perceived by him and others, but at first it had virtually no consequences. Luther then realized, however, that a dirty game was being played with men's religious convictions by the sale of indulgences, and he conceived it absolutely necessary to secure a remedy on that point. Luther never had any intention of altering religion or even of splitting the Church in any way. At first he devoted all his efforts to settling this one problem of indulgences, and from this the Reformation followed with manifest historical inevitability.

But why should it be an error to demand the overthrow of everything existing, if after that a revolution still takes place? The answer follows almost automatically from what we have said: because we thereby run the risk of wanting uncritically to change things, even where change is made impossible for all time by the laws of nature. In natural science it is only fools and visionaries, such as the inventors of perpetual-motion machines, who endeavor simply to disregard the existing laws of nature, and naturally nothing whatever comes of their efforts. Only a person who tries to alter as *little* as possible can achieve success, because he thereby makes evident what the facts compel. The small changes he eventually shows to be absolutely necessary may afterward, in the course of years or decades, enforce a change in the pattern of thought, and hence a shifting of the foundations.

I have offered you this analysis of the historical development of our subject, physics, because I am concerned that the contemporary vogue word "revolution" can lead into all sorts of mistaken courses, which a glance at the history of modern physics might help us to avoid. But, as I have said, I leave it to you to consider how far it is proper to compare revolutions in science with revolutions in society. Such an analogy can never be more than half correct, and I have only dwelt upon it here in order to give occasion for thought.

XIII

The Meaning of Beauty in the Exact Sciences

When a representative of natural science is called upon to address a meeting of the Academy of Fine Arts, he can scarcely venture to express opinions on the subject of art; because the arts are certainly remote from his own field of activity. But perhaps he may be allowed to tackle the problem of beauty. For although the epithet "beautiful" (or "fine") is indeed employed here to characterize the arts, the realm of the beautiful stretches far beyond their territory. It assuredly encompasses other regions of mental life as well; and the beauty of nature is also reflected in the beauty of natural science.

Perhaps it will be best if, without any initial attempt at a philosophical analysis of the concept of "beauty," we simply ask where we can meet the beautiful in the sphere of exact science. Here I may perhaps be allowed to begin with a personal experience. When, as a small boy, I was attending the lowest classes of the Max-Gymnasium here in Munich, I became interested in numbers. It gave me pleasure to get to know their properties, to find out, for example, whether they were prime numbers or not, and to test whether they could perhaps be represented as sums of squares, or eventually to prove that there must be infinitely many primes. Now since my father thought my

Address delivered to the Bavarian Academy of Fine Arts, Munich 1970. First published in a bibliophile edition (German and English) in the Belser-Presse collection *Meilensteine des Denkens und Forschens,* Stuttgart 1971.

knowledge of Latin to be much more important than my numerical interests, he brought home to me one day from the National Library a treatise written in Latin by the mathematician Leopold Kronecker, in which the properties of whole numbers were set in relation to the geometrical problem of dividing a circle into a number of equal parts. How my father happened to light on this particular investigation from the middle of the last century I do not know. But the study of Kronecker's work made a deep impression on me. I sensed a quite immediate beauty in the fact that, from the problem of partitioning a circle, whose simplest cases were of course familiar to us in school, it was possible to learn something about the totally different sort of questions involved in elementary number theory. Far in the distance, no doubt, there already floated the question whether whole numbers and geometrical forms exist, i.e., whether they are there outside the human mind or whether they have merely been created by this mind as instruments for understanding the world. But at that time I was not yet able to think about such problems. The impression of something very beautiful was, however, perfectly direct; it required no justification or explanation.

But what was beautiful here? Even in antiquity there were two definitions of beauty, which stood in a certain opposition to one another. The controversy between them played a great part especially during the Renaissance. The one describes beauty as the proper conformity of the parts to one another, and to the whole. The other, stemming from Plotinus, describes it, without any reference to parts, as the translucence of the eternal splendor of the "one" through the material phenomenon. In our mathematical example we shall have to stop short, initially, at the first definition. The parts here are the properties of whole numbers and laws of geometrical constructions, while the whole is obviously the underlying system of mathematical axioms to which arithmetic and Euclidean geometry belong—the great structure of interconnection guaranteed by the consistency of the axiom system. We perceive that the individual parts fit together, that as parts they do indeed belong to this whole, and without any reflection we feel the completeness and simplicity of this axiom system to be beautiful. Beauty is therefore involved with the age-old problem of the "one" and the "many" which occupied—in close connection with the problem of "being" and "becoming"—a central position in early Greek philosophy.

Since the roots of exact science are also to be found at this very point, it will be as well to retrace in broad outline the currents of thought in that early age. At the starting point of the Greek philosophy of nature there stands the question of a basic principle, from which the colorful variety of phenomena can be explained. However strangely it may strike us, the well-known answer of Thales—"Water is the material first principle of all things"—contains, according to Nietzsche, three basic philosophical demands which were to become important in the developments that followed: first, that one should seek for such a unitary basic principle; second, that the answer should be given only rationally, that is, not by reference to a myth; and third and finally, that in this context the material aspect of the world must play a deciding role. Behind these demands there stands, of course, the unspoken recognition that understanding can never mean anything more than the perception of connections, i.e., unitary features or marks of affinity in the manifold.

But if such a unitary principle of all things exists, then—and this was the next step along this line of thought—one is straightway brought up against the question how it can serve to account for the fact of change. The difficulty is particularly apparent in the celebrated paradox of Parmenides. Only being is; non-being is not. But if only being is, there cannot be anything outside this being that articulates it or could bring about changes. Hence being will have to be conceived as eternal, uniform and unlimited in space and time. The changes we experience can thus be only an illusion.

Greek thought could not stay with this paradox for long. The eternal flux of appearances was immediately given, and the problem was to explain it. In attempting to overcome the difficulty, various philosophers struck out in different directions. One road led to the atomic theory of Democritus. In addition to being, non-being can still exist as a possibility, namely as the possibility for movement and form, or in other words, as empty space. Being is repeatable, and thus we arrive at the picture of atoms in the void—the picture that has since become infinitely fruitful as a foundation for natural science. But of this road we shall say no more just now. Our purpose, rather, is to present in more detail the other road, which led to Plato's Ideas, and which carries us directly into the problem of beauty.

This road begins in the school of Pythagoras. It is there that the notion is said to have originated that mathematics, the mathematical order, was the basic principle whereby the multiplicity of phenomena could be accounted for. Of Pythagoras himself we know little. His disciples seem, in fact, to have been a religious sect, and only the doctrine of transmigration and the laying down of certain moral and religious rules and prohibitions can be traced with any certainty to Pythagoras. But among these disciples—and this was what mattered subsequently—a preoccupation with music and mathematics played an important role. Here it was that Pythagoras is said to have made the famous discovery that vibrating strings under equal tension sound together in harmony if their lengths are in a simple numerical ratio. The mathematical structure, namely the numerical ratio as a source of harmony, was certainly one of the most momentous discoveries in the history of mankind. The harmonious concord of two strings yields a beautiful sound. Owing to the discomfort caused by beat-effects, the human ear finds dissonance disturbing, but consonance, the peace of harmony, it finds beautiful. Thus the mathematical relation was also the source of beauty.

Beauty, so the first of our ancient definitions ran, is the proper conformity of the parts to one another and to the whole. The parts here are the individual notes, while the whole is the harmonious sound. The mathematical relation can therefore assemble two initially independent parts into a whole, and so produce beauty. This discovery effected a breakthrough, in Pythagorean doctrine, to entirely new forms of thought, and so brought it about that the ultimate basis of all being was no longer envisaged as a sensory material—such as water, in Thales—but as an ideal principle of form. This was to state a basic idea which later provided the foundation for all exact science. Aristotle, in his *Metaphysics,* reports that the Pythagoreans, ". . . who were the first to take up mathematics, not only advanced this study, but also having been brought up in it they thought its principles were the principles of all things. . . . Since, again, they saw that the modifications and the ratios of the musical scales were expressible in numbers; since, then, all other things seemed in their whole nature to be modelled on numbers; and numbers seemed to be the first things in the whole of nature, they supposed the elements of numbers to be the elements of all things, and the whole heaven to be a musical scale and a number." (I, 5, 985b–986a; Ross's translation.)

Understanding of the colorful multiplicity of the phenomena was thus to come about by recognizing in them unitary principles of form, which can be expressed in the language of mathematics. By this, too, a close connection was established between the intelligible and the beautiful. For if the beautiful is conceived as a conformity of the parts to one another and to the whole, and if, on the other hand, all understanding is first made possible by means of this formal connection, the experience of the beautiful becomes virtually identical with the experience of connections either understood or at least guessed at.

The next step along this road was taken by Plato, with the formulation of his theory of Ideas. Plato contrasts the imperfect shapes of the corporeal world of the senses with the perfect forms of mathematics; the imperfectly circular orbits of the stars, say, with the perfection of the mathematically defined circle. Material things are the copies, the shadow images, of ideal shapes in reality; moreover, as we should be tempted to continue nowadays, these ideal shapes are actual because and insofar as they become "act"-ive in material events. Plato thus distinguishes here with complete clarity a corporeal being accessible to the senses and a purely ideal being apprehensible not by the senses but only through acts of mind. Nor is this ideal being in any way in need of man's thought in order to be brought forth by him. On the contrary, it is the true being, of which the corporeal world and human thinking are mere reproductions. As their name already indicates, the apprehension of Ideas by the human mind is more an artistic intuiting, a half-conscious intimation, than a knowledge conveyed by the understanding. It is a reminiscence of forms that were already implanted in this soul before its existence on earth. The central Idea is that of the Beautiful and the Good, in which the divine becomes visible and at sight of which the wings of the soul begin to grow. A passage in the *Phaedrus* (251 ff.) expresses the following thought: the soul is awe-stricken and shudders at the sight of the beautiful, for it feels that something is evoked in it that was not imparted to it from without by the senses but has always been already laid down there in a deeply unconscious region.

But let us come back once more to understanding, and thus to natural science. The colorful multiplicity of the phenomena can be understood, according to Pythagoras and Plato, because and insofar as it is underlain by unitary principles of form susceptible of mathematical representation. This postulate already constitutes an anticipation of the entire program of contemporary exact science. It could not, however, be carried through in antiquity, since an empirical knowledge of the details of natural processes was largely lacking.

The first attempt to penetrate into these details was undertaken, as we know, in the philosophy of Aristotle. But in view of the infinite wealth initially presented here to the observing student of nature and the total lack of any sort of viewpoint from which an order might have been discernible, the unitary principles of form sought by Pythagoras and Plato were obliged to give place to the description of details. Thus there arose the conflict that has continued to this day in the debates, for example, between experimental and theoretical physics; the conflict between the empiricist, who by careful and scrupulous detailed investigation first furnishes the presuppositions for an understanding of nature, and the theoretician, who creates mathematical pictures whereby he seeks to order and so to understand nature—mathematical pictures that prove themselves, not only by their correct depiction of experience, but also and more especially by their simplicity and beauty, to be the true Ideas underlying the course of nature.

Aristotle himself, as an empiricist, was critical of the Pythagoreans, who, he said (*De Caelo,* II, 13, 293a), "are not seeking for theories and causes to account for observed facts, but rather forcing their observations and trying to accommodate them to certain theories and opinions of their own" and were thus setting up, one might say, as joint organizers of the universe. If we look back on the history of the exact sciences, it can perhaps be asserted that the correct representation of natural phenomena has evolved from this very tension between the two opposing views. Pure mathematical speculation becomes unfruitful because from playing with the wealth of possible forms it no longer finds its way back to the small number of forms according to which nature is actually constructed. And pure empiricism becomes unfruitful because it eventually bogs down in endless tabulation without inner connection. Only from the tension, the interplay between the wealth of facts and the mathematical forms that may possibly be appropriate to them, can decisive advances spring.

But in antiquity this tension was no longer acceptable, and thus the road to knowledge diverged for a long time from the road to the beautiful. The significance of the beautiful for the understanding of nature became clearly visible again only at the beginning of the modern period, once the way back had been found from Aristotle to Plato. And only through this change of course did the full fruitfulness become apparent of the mode of thought inaugurated by Pythagoras and Plato.

This is most clearly shown in the celebrated experiments on

falling bodies that Galileo probably did not, in fact, conduct from the leaning tower of Pisa. Galileo begins with careful observations, paying no attention to the authority of Aristotle, but, following the teaching of Pythagoras and Plato, he does try to find mathematical forms corresponding to the facts obtained by experiment, and thus arrives at his laws of falling bodies. However, and this is a crucial point, he is obliged, in order to recognize the beauty of mathematical forms in the phenomena, to idealize the facts, or, as Aristotle disparagingly puts it, to force them. Aristotle had taught that all moving bodies not acted upon by external forces eventually come to rest, and this was the general experience. Galileo maintains, on the contrary, that in the absence of external forces bodies continue in a state of uniform motion. Galileo could venture to force the facts in this way because he could point out that moving bodies are of course always exposed to a frictional resistance, and that motion in fact continues the longer, the more effectively the frictional forces can be cut off. In exchange for this forcing of the facts, this idealization, he obtained a simple mathematical law, and this was the beginning of modern exact science.

Some years later, Kepler succeeded in discovering new mathematical forms in the data of his very careful observations of the planetary orbits, and in formulating the three famous laws that bear his name. How close Kepler felt himself in these discoveries to the ancient arguments of Pythagoras, and how much the beauty of the connections guided him in formulating them, can be seen from the fact that he compared the revolutions of the planets about the sun with the vibrations of a string, and spoke of a harmonious concord of the different planetary orbits, of a harmony of the spheres. At the end of his work on the harmony of the universe, he broke out into this cry of joy: "I thank thee, Lord God our Creator, that thou allowest me to see the beauty in thy work of creation." Kepler was profoundly struck by the fact that here he had chanced upon a central connection which had not been conceived by man, which it had been reserved to him to recognize for the first time—a connection of the highest beauty. A few decades later, Isaac Newton in England set forth this connection in all its completeness, and described it in detail in his great work *Principia Mathematica.* The road of exact science was thus pointed out in advance for almost two centuries.

But are we dealing here with knowledge merely, or also with the beautiful? And if the beautiful is also involved, what role did it play in the discovery of these connections? Let us again recall the first definition given in antiquity: "Beauty is the proper conformity of the parts to one another and to the whole." That this criterion applies in the highest degree to a structure like Newtonian mechanics is something that scarcely needs explaining. The parts are the individual mechanical processes—those which we carefully isolate by means of apparatus no less than those which occur inextricably before our eyes in the colorful play of phenomena. And the whole is the unitary principle of form which all these processes comply with and which was mathematically established by Newton in a simple system of axioms. Unity and simplicity are not, indeed, precisely the same. But the fact that in such a theory the many are confronted with the one, that in it the many are unified, itself has the undoubted consequence that we also feel it at the same time to be simple and beautiful. The significance of the beautiful for the discovery of the true has at all times been recognized and emphasized. The Latin motto *"Simplex sigillum veri"*—"The simple is the seal of the true"—is inscribed in large letters in the physics auditorium of the University of Göttingen, as an admonition to those who would discover what is new; and another Latin motto, *"Pulchritudo splendor veritatis"*—"Beauty is the splendor of truth"—can also be interpreted to mean that the researcher first recognizes truth by this splendor, by the way it shines forth.

Twice more in the history of exact science, this shining forth of the great connection has been the crucial signal for a significant advance. I am thinking here of two events in the physics of our own century, the emergence of relativity theory and the quantum theory. In both cases, after years of vain effort at understanding, a bewildering plethora of details has been almost suddenly reduced to order by the appearance of a connection, largely unintuitable but still ultimately simple in its substance, that was immediately found convincing by virtue of its completeness and abstract beauty—convincing, that is, to all who could understand and speak such an abstract language.

But now, instead of pursuing the historical course of events any further, let us rather put the question quite directly: What is it that shines forth here? How comes it that with this shining forth of the beautiful into exact science the great connection becomes recognizable, even before it is understood in detail and before it can be rationally demonstrated? In what does the power of illumination consist, and what effect does it have on the onward progress of science?

Perhaps we should begin here by recalling a phenomenon that may be described as the unfolding of abstract structures. It can be illustrated by the example of number theory, which we referred to at the outset, but one may also point to comparable processes in the evolution of art. For the mathematical foundation of arithmetic, or the theory of numbers, a few simple axioms are sufficient, which in fact merely define exactly what counting is. But with these few axioms we have already posited that whole abundance of forms which has entered the minds of mathematicians only in the course of the long history of the subject—the theory of prime numbers, of quadratic residues, of numerical congruences, etc. One might say that the abstract structures posited in and with numbers have unfolded visibly only in the course of mathematical history, that they have generated the wealth of propositions and relationships that makes up the content of the complicated science of number theory. A similar position is also occupied—at the outset of an artistic style in architecture, say—by certain simple basic forms, such as the semicircle and rectangle in Romanesque architecture. From these basic forms there arise in the course of history new, more complicated and also altered forms, which yet can still in some way be regarded as variations on the same theme; and thus from the basic structures there emerges a new manner, a new style of building. We have the feeling, nonetheless, that the possibilities of development were already perceivable in these original forms, even at the outset; otherwise it would be scarcely comprehensible that many gifted artists should have so quickly resolved to pursue these new possibilities.

Such an unfolding of abstract basic structures has assuredly also occurred in the instances I have enumerated from the history of the exact sciences. This growth, this constant development of new branches, went on in Newtonian mechanics up to the middle of the last century. In relativity theory and the quantum theory we have experienced a similar development in the present century, and the growth has not yet come to an end.

Moreover, in science as in art, this process also has an important social and ethical aspect; for many men can take an active part in it. When a great cathedral was to be built in the Middle Ages, many master masons and craftsmen were employed. They were imbued with the idea of beauty posited by the original forms, and were compelled by their task to carry out exact and meticulous work in accordance with these forms. In similar fashion, during the two centuries following Newton's discovery, many mathematicians, physicists and technicians were called upon to deal with specific mechanical problems according to the Newtonian methods, to carry out experiments or to effect technical applications; and here, too, extreme care was always required in order to attain what was possible within the framework of Newtonian mechanics. Perhaps it may be said in general that by means of the underlying structures, in this case Newtonian mechanics, guidelines were drawn or even standards of value set up, whereby it could be objectively decided whether a given task had been well or ill discharged. It is the very fact that specific requirements have been laid down, that the individual can assist by small contributions in the attainment of large goals and that the value of his contribution can be objectively determined, which gives rise to the satisfaction proceeding from such a development for the large number of people involved. Hence even the ethical significances of technology for our present age should not be underestimated.

The development of science and technology has also produced, for example, the Idea of the airplane. The individual technician who assembles some component for such a plane, the artisan who makes it, knows that his work calls for the utmost care and exactitude and that the lives of many may well depend upon its reliability. Hence he can take pride in a well-executed piece of work, and delights, as we do, in the beauty of the aircraft, when he feels that in it the technical goal has been realized by properly adequate means. Beauty, so runs the ancient definition we have already often cited, is the proper conformity of the parts to one another and to the whole, and this requirement must also be satisfied in a good aircraft.

But in pointing thus to the evolution of beauty's ground structure, to the ethical values and demands that subsequently emerge in the historical course of development, we have not yet answered the question we asked earlier, namely, what it is that shines forth in these structures, how the great connection is recognized even before it is rationally understood in detail. Here we ought to reckon in advance with the possibility that even such recognition may be founded upon illusions. But it cannot be doubted that there actually is this perfectly immediate recognition, this shuddering before the beautiful, of which Plato speaks in the *Phaedrus*.

Among all those who have pondered on this question, it seems to have been universally agreed that this immediate recognition is not a consequence of discursive (i.e., rational) thinking. I should like here to cite two statements, one from Johannes Kepler, who has already been referred to, and the other, in our own time, from the Zürich atomic physicist Wolfgang Pauli, who was a friend of the psychologist, Carl Jung. The first passage is to be found in Kepler's *Harmony of the World:*

> That faculty which perceives and recognizes the noble proportions in what is given to the senses, and in other things situated outside itself, must be ascribed to the lower region of the soul. It lies very close to the faculty which supplies formal schemata to the senses, or deeper still, and thus adjacent to the purely vital power of the soul, which does not think discursively, i.e., in conclusions, as the philosophers do, and employs no considered method, and is thus not peculiar only to man, but also dwells in wild animals and the dear beasts of the field. . . . Now it might be asked how this faculty of the soul, which does not engage in conceptual thinking, and can therefore have no proper knowledge of harmonic relations, should be capable of recognizing what is given in the outside world. For to recognize is to compare the sense perception outside with the original pictures inside, and to judge that it conforms to them. Proclus has expressed the matter very finely in his simile of awakening, as from a dream. For just as the sensorily presented things in the outer world recall to us those which we formerly perceived in the dream, so also the mathematical relations given in sensibility call forth those intelligible archetypes which were already given inwardly beforehand, so that they now shine forth truly and vividly in the soul, where before they were only obscurely present there. But how have they come to be within? To this I answer that all pure Ideas or archetypal patterns of harmony, such as we were speaking of, are inherently present in those who are capable of apprehending them. But they are not first received into the mind by a conceptual process, being the product, rather, of a sort of instinctive intuition of pure quantity, and are innate in these individuals, just as the number of petals in a plant, say, is innate in its form principle, or the number of its seed chambers is innate in the apple.

So far Kepler. He is therefore referring us here to possibilities already to be found in the animal and plant kingdoms, to innate archetypes that bring about the recognition of forms.

In our own day, Adolf Portmann, in particular, has described such possibilities, pointing for example to specific color patterns seen in the plumage of birds, which can possess a biological meaning only if they are also perceived by other members of the same species. The perceptual capacity will therefore have to be just as innate as the pattern itself. We may also consider bird song at this point. At first the biological requirement here may well have been simply for a specific acoustic signal, serving to seek out the partner and understood by the latter. But to the extent that this immediate biological function declines in importance, a playful enlargement of the stock of forms may ensue, an unfolding of the underlying melodic structure, which is then found enchanting as song by even so alien a species as man. The capacity to recognize this play of forms must at all events be innate to the species of bird in question, for certainly it has no need of discursive, rational thought. In man, to cite another example, there is probably an inborn capacity for understanding certain basic forms of the language of gesture, and thus for deciding, say, whether the other has friendly or hostile intentions—a capacity of the utmost importance for man's communal life.

Ideas similar to those of Kepler have been put forward in an essay by Pauli. He writes:

> The process of understanding in nature, together with the joy that man feels in understanding, i.e., in becoming acquainted with new knowledge, seems therefore to rest upon a correspondence, a coming into congruence of preexistent internal images of the human psyche with external objects and their behavior. This view of natural knowledge goes back, of course, to Plato and was . . . also very plainly adopted by Kepler. The latter speaks, in fact, of Ideas, preexistent in the mind of God and imprinted accordingly upon the soul, as the image of God. These primal images, which the soul can perceive by means of an innate instinct, Kepler calls archetypes. There is very wide-ranging agreement here with

the primordial images or archetypes introduced into modern psychology by C. G. Jung, which function as instinctive patterns of ideation. In that modern psychology has given proof that all understanding is a protracted affair, accompanied by processes in the unconscious long before the content of consciousness can be rationally formulated, it has again directed attention to the preconscious, archaic stage of cognition. At this stage, the place of clear concepts is taken by images of strongly emotional content, which are not thought but are seen pictorially, as it were, before the mind's eye. Insofar as these images are the expression of a suspected but still unknown state of affairs, they can also be called symbolic, according to the definition of a symbol proposed by Jung. As ordering operators and formatives in this world of symbolic images, the archetypes function, indeed, as the desired bridge between sense perceptions and Ideas, and are therefore also a necessary precondition for the emergence of a scientific theory. Yet one must beware of displacing this a priori of knowledge into consciousness, and relating it to specific, rationally formulable Ideas.

In the further course of his inquiries, Pauli then goes on to show that Kepler did not derive his conviction of the correctness of the Copernican system primarily from any particular data of astronomical observation but rather from the agreement of the Copernican picture with an archetype which Jung calls a *mandala,* and which was also used by Kepler as a symbol for the Trinity. God, as prime mover, is seen at the center of a sphere; the world, in which the Son works, is compared with the sphere's surface; and the Holy Ghost corresponds to the beams that radiate from center to surface of the sphere. It is naturally characteristic of these primal images that they cannot really be rationally or even intuitively described.

Although Kepler may have acquired his conviction of the correctness of Copernicanism from primal images of this kind, it remains a crucial precondition for any usable scientific theory that it should subsequently stand up to empirical testing and rational analysis. In this respect the sciences are in a happier position than the arts, since for science there is an inexorable and irrevocable criterion of value that no piece of work can evade. The Copernican system, the Keplerian laws and the Newtonian mechanics have subsequently proved themselves in the interpreting of phenomena, in observational findings and in technology, over such a range and with such extreme accuracy that after Newton's *Principia* it was no longer possible to doubt that they were correct. Yet even here there was still an idealization involved, such as Plato had held necessary and Aristotle had disapproved.

This only came out in full clarity some fifty years ago, when it was realized from the findings in atomic physics that the Newtonian scheme of concepts was no longer adequate to cope with the mechanical phenomena in the interior of the atom. Since Planck's discovery of the quantum of action, in 1900, a state of confusion had arisen in physics. The old rules, whereby nature had been successfully described for more than two centuries, would no longer fit the new findings. But even these findings were themselves inherently contradictory. A hypothesis that proved itself in one experiment failed in another. The beauty and completeness of the old physics seemed destroyed, without anyone having been able, from the often disparate experiments, to gain a real insight into new and different sorts of connection. I don't know if it is fitting to compare the state of physics in those twenty-five years after Planck's discovery (which I too encountered as a young student) to the circumstances of contemporary modern art. But I have to confess that this comparison repeatedly comes to my mind. The helplessness when faced with the question of what to do about the bewildering phenomena, the lamenting over lost connections, which still continue to look so very convincing—all these discontents have shaped the face of both disciplines and both periods, different as they are, in a similar manner. We are obviously concerned here with a necessary intervening stage, which cannot be by-passed and which is preparing for developments to come. For as Pauli told us, all understanding is a protracted affair, inaugurated by processes in the unconscious long before the content of consciousness can be rationally formulated. The archetypes function as the desired bridge between the sense perceptions and the Ideas.

At that moment, however, when the true Ideas rise up, there occurs in the soul of him who sees them an altogether indescribable process of the highest intensity. It is the amazed awe that Plato speaks of in the *Phaedrus,* with which the soul remembers, as it were, something it had unconsciously possessed all along. Kepler says: *"Geometria est archetypus pulchritudinis mundi";* or, if we may translate in more general terms—"Mathematics is the archetype of the beauty of the world." In atomic physics this process took place not quite fifty years ago, and has again restored exact science, under entirely new presuppositions, to that state of harmonious completeness which for a quarter of a century it had lost. I see no reason why the same thing should not also happen one day in art. But it must be added, by way of warning, that such a thing cannot be made to happen—it has to occur on its own.

Ladies and gentlemen, I have set this aspect of exact science before you because in it the affinity with the fine arts becomes most plainly visible, and because here one may counter the misapprehension that natural science and technology are concerned solely with precise observation and rational, discursive thought. To be sure, this rational thinking and careful measurement belong to the scientist's work, just as the hammer and chisel belong to the work of the sculptor. But in both cases they are merely the tools and not the content of the work.

Perhaps at the very end I may remind you once more of the second definition of the concept of beauty, which stems from Plotinus and in which no more is heard of the parts and the whole: "Beauty is the translucence, through the material phenomenon, of the eternal splendor of the 'one.'" There are important periods of art in which this definition is more appropriate than the first, and to such periods we often look longingly back. But in our own time it is hard to speak of beauty from this aspect, and perhaps it is a good rule to adhere to the custom of the age one has to live in, and to keep silent about that which it is difficult to say. In actual fact the two definitions are not so very widely removed from one another. So let us be content with the first and more sober definition of beauty, which certainly is also realized in natural science; and let us declare that in exact science, no less than in the arts, it is the most important source of illumination and clarity.

XIV

The End of Physics?

The physics of elementary particles stands in the center of physical interest nowadays, and sometimes it has been asked whether, with the solution of the problems posed in this area, we should simultaneously have come to the end of physics as a whole. For, it might be argued, all matter and all radiation consist of elementary particles; hence a complete knowledge of the laws governing their properties and behavior, in the shape, say, of a "world formula," would also be bound to establish the basic framework for all physical processes. So even if extended developments could still be appended in applied physics and technology, the questions of principle would all have been settled, and fundamental research in physics would have come to an end.

This thesis of a possible completion to physics is contradicted by the experience of earlier periods, in which it was also wrongly supposed that physics would soon be concluded. Max Planck has recorded that his teacher, Jolly, advised him against the study of physics, since after all it was essentially finished with, so that for anyone who wanted to do active scientific research it would scarcely be worthwhile to go into this field. Nobody would wish to make any such false prediction nowadays, and we ought therefore to ask whether anywhere in the past history of physics there have been at least subareas in which a final formulation of nature's laws has been arrived at; in

Article in the *Süddeutsche Zeitung,* October 6, 1970.

which we can therefore be confident that a thousand or millions of years hence, or in any star systems however remote, the phenomena will continue precisely in accordance with the same mathematically formulable laws.

Such closed-off subareas undoubtedly exist. To pick out a very special example: the laws of the lever were formulated by Archimedes more than two thousand years ago; we can have no doubt that, at all times and places, they retain their validity. Thus if moon travelers use levers for their work on the moon, they self-evidently and successfully assume the old laws of Archimedes to be correct. The same seems also to hold good for the whole of Newtonian mechanics. The moon travelers rely unhesitatingly on its statements, and act accordingly. But at this point already an objection might be raised: has not the Newtonian mechanics been improved by means of relativity theory and the quantum theory? When it comes to high degrees of accuracy, must not the moon travelers take these refinements into account? And if they have to do this, don't the improvements show that even mechanics is still far from being a finished science?

To find an answer here, we must first of all point out that the great comprehensive formulations of natural laws, such as first became possible in Newtonian mechanics, are concerned with idealizations of reality, not with reality itself. The idealization comes about through the fact that we approach reality with certain concepts, which have proved themselves in the description of the phenomena and which thereby give the latter a certain aspect. In mechanics, for example, we make use of such concepts as position, time, velocity, mass and force. But by this means we restrict the picture of reality—or if you will, we stylize it—since we simultaneously forgo all those features in the phenomena which cannot be captured by means of these concepts. If we remain aware of these restrictions, it is possible to claim that mechanics is brought to completion in the Newtonian theory, meaning that mechanical phenomena, so far as they can be described at all in the concepts of Newtonian physics, also take place strictly in accordance with the laws of that physics. As I have said, we are convinced that these statements will still hold good millions of years from now and in the remotest star systems, and we believe that within the framework of its concepts Newtonian physics cannot be improved upon. But we are by no means able to claim that all phenomena can be described in terms of these concepts.

With the reservations mentioned, it is therefore possible to say that Newtonian mechanics is a completed theory. Such a closed-off theory is characterized by a system of definitions and axioms that establishes the fundamental concepts and their interrelations; and also by the requirement that there is a wide realm of experiences, of observable phenomena, that can be described with high accuracy by means of this system. The theory is then the idealization, valid for all time, of this realm of experience.

But there are other realms of experience, and hence other closed-off theories as well. In the nineteenth century the theory of heat, in particular, took on final form, in this sense, as a statistical statement about systems with very many degrees of freedom. The fundamental axioms of this theory define and connect such concepts as temperature, entropy and energy, of which the first two, temperature and entropy, make no appearance whatever in Newtonian mechanics, while the last, energy, plays an important role in every field of experience and not merely in mechanics. Since the work of Willard Gibbs, the statistical theory of heat can likewise be reckoned a final and closed-off theory, nor can we doubt that its laws apply everywhere, at all times, with the highest accuracy—although naturally only to those phenomena which can be dealt with by means of such concepts as temperature, entropy and energy. This theory, too, is an idealization; and we know that there are many conditions, e.g., of matter in the gaseous state, where one cannot speak of temperature and so cannot apply the laws of this heat theory either.

From what has been said it will already be clear that in

physics, at all events, there do exist closed-off theories, which can be regarded as idealizations for restricted fields of experience and which claim to be valid for all time. But there can obviously be no talk here, as yet, of any closing off of physics as a whole.

In the last two hundred years, quite new fields of experience have been opened up by experiment. Since the foundational inquiries of Luigi Galvani and Alessandro Volta, the phenomena of electromagnetism have been studied with ever greater exactness, their relationships to chemistry being demonstrated by Faraday and those to optics by Heinrich Hertz. The fundamental facts of atomic physics were first disclosed by findings in chemistry, and then explored in every detail by experiments in electrolysis, in discharge processes in gases and later in radioactivity. For an understanding of this gigantic new territory, the closed-off theories of an earlier day were inadequate. And so new and more comprehensive theories were framed, which can be regarded as idealizations of these new regions of experience. The theory of relativity emerged from the electrodynamics of moving bodies and has led to new insights into the structure of space and time. The quantum theory gives an account of the mechanical processes in the interior of the atom, but it also incorporates Newtonian mechanics, as the limiting case in which we are able to objectify the events completely and can neglect the interaction between the object under investigation and the observer himself.

Relativity theory, no less than quantum mechanics, can also be viewed as a closed-off theory, a very comprehensive idealization of exceedingly large tracts of experience, of whose laws we can take it that they are valid everywhere and at all times—but again only for those areas of experience which can be apprehended by means of these concepts.

Finally, in recent decades, the physics of elementary particles has been opened up by investigations on cosmic radiation, and especially through experiments carried out with the aid of large accelerators (e.g., at Berkeley, Geneva, Brookhaven and Serpuchov). In the process, novel features have made their appearance in the phenomena, which have displayed the old problem of the smallest parts of matter in a new light. In the past development of physics it had repeatedly been found that the structures which had initially been regarded as the smallest material parts could be divided, by the application of even larger forces, into pieces smaller still. The chemist's atoms could not indeed be divided by chemical means; but in discharge tubes, under the influence of stronger electrical forces, they certainly could be split up into atomic nuclei and their surrounding electrons. On collision with energetic atomic nuclei, these nuclei themselves could be divided still further, and it was recognized that all atomic nuclei are made up of two basic building blocks, protons (hydrogen nuclei) and neutrons, which were designated along with the electrons as elementary particles.

It was therefore natural to suppose that even protons and neutrons could be broken up still further if still larger forces were employed—if they were hurled at each other, for example, with extraordinarily high energies. This was actually tried out in the great accelerators, but it turned out that in such collisions something different occurs. The high energy of motion of the elementary particles shot at one another is transformed into matter, that is, new elementary particles arise out of the collision, but these are apt to be in no way smaller than the particles that were made to collide. Thus we cannot really speak any longer of a "division" here. With the elementary particles that we know today, and with the great accelerators that we use to experiment on them, we have thus arrived at the boundary where the concept of division loses its meaning, and hence can assume with a good conscience that the elementary particles now known are really the smallest parts of matter, so far as we can give this concept any meaning at all.

This new area of experience, the physics of elementary particles, could not be represented in terms of the closed-off theories —quantum mechanics and relativity theory—that had previously been developed, although in both these theories very comprehensive idealizations were already involved. But quantum mechanics continued, no less than did the old Newtonian mechanics, to presuppose the existence of immutable mass points; nothing is said in it of a transformation of energy into matter. Relativity theory, on the other hand, neglects those features of nature which are tied up with Planck's quantum of action; it therefore continues to presuppose that the phenomena can be objectified in the manner of classical physics. For the physics of elementary particles it has therefore been necessary to seek out a new and still more comprehensive idealization, which should include both relativity theory and quantum theory as limiting cases, and which should account for the complicated spectrum of the elementary particles, much as quantum mechanics has been able to account, say, for the complex optical spectrum of the iron atom. There can be no doubt that this idealization will one day be represented in mathematical terms; but whether the form so far proposed for this mathematical representation is already adequate will be verified only after further experimental and theoretical investigation. Independently of this latter problem, which calls for no discussion here, it can nevertheless be asked whether, once this idealization is effected, physics will have been brought to completion. Since all physical objects are made up of elementary particles, it might be inferred that a complete knowledge of the laws governing the behavior of these particles would be equivalent to a complete knowledge of the laws of behavior of all physical objects, and to that extent it might then be said that physics had come to an end.

Such an inference would be inadmissible, however, since it fails to give sufficient attention to an important point. Even a closed-off theory of elementary particles—whether we now choose to call it a "world formula" or not—must still be regarded as an idealization. To be sure, it gives an exact picture of an enormously wide range of phenomena, but there may still be other phenomena that the concepts of this idealization are unable to capture. The most conspicuous proof of this possibility is biology. Admittedly, all biological objects consist of elementary particles; but the concepts we use to describe biological processes, e.g., the concept of life itself, do not appear in our idealization; so there must be still further developments of physics in this direction. At best it might be objected here that it would then be no longer a matter of physics but simply of biology, and that physics could thus still be closed off. But the boundaries between physics and the neighboring sciences are so fluid that little would be gained by distinctions of that kind. The majority of physicists are therefore agreed that, precisely in view of its undefined boundaries with neighboring areas, it would not be proper to speak of a closure of physics.

Some physicists would even deny, however, that any closing off of the narrower region of elementary-particle physics is to be expected in the foreseeable future. It is pointed out that with the building of ever larger accelerators we are pushing forward to ever higher energies in the colliding elementary particles, and that in the process a new region, still unknown, may be revealed. But this view rests on a supposition without warrant in either experience or theory, namely that with a further increase in the energies, qualitatively new phenomena are bound to appear. In cosmic radiation, where the energies of the colliding particles can be up to a thousand times greater than in the largest accelerators yet built, no such qualitatively new phenomena have been found. Nor has there been any discovery of the "quark" particles hypothetically assumed by many theoreticians. There are therefore neither experimental nor theoretical grounds for this new and unknown region, though its existence cannot be ruled out.

So long as no such new experimental regions make their appearance, the question of closing off physics will have to be thought of primarily in terms of the fluid boundaries with neighboring areas and of the different sorts of conceptual schemes that find application there. Nor is it by any means merely a matter of the natural sciences. These adjacent territories also include mathematics, information theory and philosophy, and in time to come it will often be difficult, perhaps, to decide whether an advance in knowledge represents a step forward in physics, information theory, or philosophy, whether physics is expanding into biology or whether biology is employing physical methods and approaches to an ever greater extent. It would thus be possible to speak of a closing off of physics only if we were arbitrarily prepared to define certain methods and conceptual patterns as physical ones and to assign other ways of putting the problem to other sciences. But this is hardly likely to happen; for the characteristic feature of the coming development will surely consist in the unification of science, the conquest of the boundaries that have grown up historically between the different individual disciplines.

XV

Science in the Contemporary University

The Quincentenary Celebrations of the University of Munich should be taken as an occasion for considering the role of science in the contemporary university, or better, in the shaping of the university today. When a university can look back on five hundred years of history, when it is recognized without reservation, in its historic progress from Ingolstadt by way of Landshut to Munich, as the same institution, this can be possible only because, with all its preservation of tradition, it has changed in the course of centuries, repeatedly adapting itself to the life of the time, or, as we would be more apt to say nowadays, to the current state of society. During the last 150 years the strongest forces for change have arisen from the interplay of science and technology. Particularly since the liberation of atomic energy some twenty-five years ago, technology has overflowed all its previous bounds, and it is therefore natural to inquire into the effects of these gigantic forces on the contemporary university. But before I do this, I must offer an apology. Since the last war I have been further removed than formerly from university life. I have not taken an especially active part in this life, and have therefore not experienced directly the trials and tribulations of the university. I am therefore obliged to contemplate it from a certain distance; but perhaps from afar off the proportions, the relative importance of particular happenings, can be better

Official Address at the Quincentenary Celebrations of the University of Munich, 1972.

discerned than from close at hand. At all events, I must try to proceed from such a viewpoint in the observations that follow.

We shall first of all have to deal here with the direct effects produced on university life by the fact that modern science and technology have got to be taught, that research in these areas is largely carried on in the university and that in virtue of these new areas the equipping of the university has become incomparably more expensive. But after that we shall also have something to say of the profounder modifications now occurring at the university owing to the changes that have taken place in the society, or political landscape, in which it is embedded. These consequences also include, for example, the crisis phenomena of the past five years, evoked by the quarrel of the student body with the world of science and technology and its social structure. Finally we must pose the central question whether it is still possible, in this scientific and technical world, to strive for the freedom and breadth of thought whose transmission has always seemed to us one of the university's most important tasks.

Let us begin, therefore, with the expansion of scientific studies at the university. In the early days of universities, the sciences belonged to the *artes liberales* and played only a subordinate role. There were four faculties, of theology, jurisprudence, medicine and these same *artes liberales,* which at that time included philosophy, philology, natural science and much else. The profitable sciences, the *scientiae lucrativae,* as they were then called, were jurisprudence and medicine, and those educated in those sciences were most urgently needed in society. Not until the sixteenth century was the faculty of arts transformed into the philosophical faculty; the conflict began between the old and the new sciences, and in the seventeenth century chemistry and botany were already playing an important part, in virtue of their close connection with medicine. In the eighteenth century the Enlightenment opened up an increasingly large area to empirical and rational inquiry, and with the start of the nineteenth century there began that victorious advance of pragmatic science, arising from the interplay of science and technology, which ever since has wrought a fundamental transformation in the life of society itself. In the second half of the nineteenth century there came into being the colleges of advanced technology, in which the scientific methods of modern technology were taught and further developed. But not until 1937 was the old-style philosophy faculty at Munich divided into the faculties of arts and natural science. In recent years the science faculty has again become so large and unwieldy that it has been split up into fields of study possessing a certain independence. In the course of their history, therefore, the sciences have acquired an ever-increasing status in the university, and it is inevitable that under their influence changes have had to occur in the intellectual structure of the university itself, in the mode of work in its branches and in the organization of studies. We are still in the midst of those changes, and nobody knows what the eventual pattern of the university will look like—if indeed there is to be any such pattern.

Among the first things to happen in this connection has been the renewed outbreak of an old quarrel that already loomed large during the period of the Enlightenment. Should the university—so the question runs—be viewed merely as an assemblage of a series of separate disciplines, in which education is carried on in rigidly ordered courses of study, and thus largely in a pattern of separate schools, or should a broader scientific education provide a conspectus of the various sciences and the linkage between disciplines? Should it instill a habit of critical scientific thought, a conscious moral purpose trained in an understanding of principles, which may then subsequently find application when embarking on a career? In the second case we shall allow the student much freedom in choosing his subjects and course of study; he should ultimately decide what seems to him worth knowing and learning entirely on his own. Wilhelm von Humboldt, in his university reforms at the beginning of the nineteenth century, undoubtedly had this second view in mind; even today it still lingers as a desirable ideal before many who are shaping, and working in, the modern university. But even in those days the university was not entirely able to break away from the assemblage of departments, and there will be equally little chance for it to do so now. For society demands in the first instance trained specialists, lawyers, doctors, teachers, and in our own day chemists, physicists and mathematicians. The more complex the economic and social structure that supports the life of society, the more urgent the need for well-trained professional men; the more important, however, does it also become to have some people able to look out beyond the narrowly professional sphere.

On one point only has the preponderance of modern science come closer to the Humboldtian view: research, and with it scientific thinking, plays a larger part nowadays than knowledge and erudition. In view of the illimitable abundance of possible knowledge, we have grown accustomed to the idea that it can be stored up in libraries. It is only scientific thinking, the method that leads to new knowledge, the insight into possible sources of error, the care in preparing a line of argument, whose study and practice are seen as the true task of the university. The enormous economic and political consequences of the expansion in science have accordingly brought it about that the researcher and discoverer enjoys a higher esteem nowadays than the man of learning. They have surrounded research work with a halo that is not always helpful to the cause of shaping the university. For just as Humboldt's view eventuated in a two-class system of education, a separation of the academically educated from the people, so overestimation of successful research work promotes a form of class-thinking within science, which through one-sided standards can provoke discord among those who should be maintaining the spirit of the university. With so high a value put upon research, it is too easily overlooked that an academic teacher who produces a large number of outstanding students may be a more useful member of society than another who publishes numerous contributions to research. But the capacity for critical thinking in science will in one way or another play a more important part than copious knowledge.

The expansion of science, the opening up of so many previously unknown areas and the elaboration of innumerable new methods, has also led, during the last fifty years, to another lamentable deformity in the German university, the combating and reversal of which will be one of the most important tasks in the coming years. I refer to the lengthening of the period of graduate work. Whereas the student of fifty years ago could still generally terminate his studies at the end of the eighth semester, that is, the fourth academic year, with the doctoral examination or teacher's professional examination, it is not infrequent nowadays to demand a study period of eight to ten years before finishing the doctoral examination. Max Planck took his doctorate—though that was now almost a hundred years ago—at the age of twenty-one, and even fifty years ago such early graduations were not so very rare. Nowadays the young academic often starts his professional career only at twenty-eight or thirty.

In justification, there is an appeal to the constant enlargement of the material to be known, the complexity of modern science, which does not permit the young physicist, chemist or physician to be released into professional life after only a four-year course of study. Without—so it is alleged—a considerable lowering of standards, an abridgment of the curriculum is not possible. Perhaps the thought also enters, on occasion, that the professor's research work can only be carried on with well-trained collaborators, and that any assistants other than doctoral candidates, who already have a lengthy period of study behind them, could scarcely be found. But all these arguments are controverted by the simple requirement that a young man, who at twenty-five is at the height of his physical and mental activity, should be taking responsibility, should be standing on his own feet in society and can no longer sit at his school desk and work in a purely receptive manner. This requirement, in my view, cannot be set aside. We should also bear in mind that even the most thorough training cannot suffice to provide adequate knowledge for the entire remainder of a man's professional life. For science and technology change so rapidly that further professional training at a later stage is indispensable. Above all, however, it must be pointed out that the overextended course of study represents a specifically German malady. This may be due to our exaggerated propensity for thoroughness, our lack of talent for finding reasonable compromises or our striving for perfection. At all events, the arrangements at universities in other countries, in the West as in the East, in Britain and America as in Russia and China, show that essentially shorter courses of graduate study are possible. We must therefore learn here from the others as quickly as we can; a curricular reform on this point is an absolute necessity.

Here we enter upon another problem, the relationship of the university to society and government. The growing importance of science and technology means that this relation must inevitably become closer than in the past. Even in earlier times, to be sure, the princes drew their counselors from the academically educated class, the law required good lawyers and the practice of medicine capable doctors. But the universities could still lead a special existence in an area largely secluded from society; autonomy and self-government provided for a sufficient remoteness, and the modest requirements necessary for thought and research often made possible an undisturbed life in the ivory tower of learning. From the beginning of the nineteenth century onward, however, the significance for the life of the community of the results achieved in the university becomes apparent everywhere: geology furnishes the scientific basis for mining, chemistry provides for an enormous intensification of agriculture, lighting and communication techniques rely upon advances in physics. The history of the University of Munich is rich in achievements even in the practically applied sciences. The immediate consequence of this development is the better equipment of the universities by the state and also at times by industry. Modern experimental research calls at many points for complicated and therefore costly apparatus; the budgets of the university institutes were therefore bound to grow. But with all institutional budgets growing at an equal pace, the financial limits would have been reached so rapidly that for those departments in which costly apparatus was employed the improvement would not have been enough.

Thus two ways out have been tried. The first consists in the founding of special, highly endowed institutes for research, which retain only a loose connection with the university or are totally separated from it; here I may instance the institutes of the Max Planck Society. The second consists in the formation of points of concentration within the university, in the creation of special fields of research. In both cases it is the universality of the university that is affected. It is no longer possible to pursue every science, in every university, with the most modern means of research. A very dubious consequence of this development is the striving for ever larger institutional funding that many institute directors engage in almost as a matter of course. How justified are these demands? It is true enough that much research work can be carried on only with very large resources. If one opts for such a line of research, one must therefore have ample funds. It will also have to be granted, however, that even if the institutional budget is a modest one, it ought to be possible to find research projects which can be carried on with these modest funds and whose results may perhaps be of more importance than those obtained with the expensive apparatus. In the selection of research projects we should therefore be even more circumspect than formerly, because we are already under a further obligation to consider whether the other requirements of society permit such large expenditures for scientific research.

The example of America has much to teach us here, but we should not uncritically imitate everything that goes on there. In my own science, for example, the impression has taken root there that difficult problems can be solved simply by measures of organization. Billions of dollars are made available, thousands of physicists are trained and costly giant accelerators are built. The landscape of research is flattened out, as it were, with a steamroller, in the specific expectation that it will then become accessible to everybody. But if we look more carefully to see where the important successes of American physics have been obtained in the last ten years, we perceive that a number of these successes, perhaps the most important, are due to outsiders who have gone their own way with limited resources; who—to use an Americanism—have not been riding the bandwagon but who constantly, over a period of many years off the broad highway, have secured unexpected research results. As an example, I could mention Joe Weber's investigations on gravitational waves.

Large-scale government funding for university research presents yet another aspect that calls for serious consideration, namely the increasing influence of the state upon the university. If society has to provide large sums for research, it is inevitable that it should also require the application of these sums to be publicly controlled for the benefit of society, and indeed for the benefit of those social goals which society itself considers important. Here the danger arises that the technical applications of research results may become the chief measure of value and that fundamental research may suffer accordingly. We perceive this consequence most clearly in the totalitarian states, where official control of the universities is exceptionally rigid. In China, for example, Mao Tse-tung in one of his well-known pronouncements has declared the machine-tool factory in Shanghai a model for the way in which the universities ought to be training technicians from the working class. We should not care to see the university yoked so directly to the needs of industry, and we believe that basic research, performed, that is, for the sake of pure knowledge, is a highly important constituent of the scientific life, even in the interests of society. In China, moreover, during the same period immediately after the Cultural Revolution, the time given to university studies was drastically curtailed, on the plea that the long period of study was encouraging a false habit of elite thinking among the students. An official article in the Peking *People's Daily* stated that there was a danger that the students would look down upon the workers and peasants and regard themselves as great people, and that it was necessary to advise them to discard their affected ways.

In the states where a parliamentary democracy provides for a relatively free form of government, there has been a strengthening not only of state influence upon the university but also, conversely, of the influence of scientific circles upon the research and educational policies of the state. Consultative committees have everywhere been formed, in which representatives of officialdom and science confer about the distribution of research funds and discuss the guidelines for government research and educational policy. Granted, this is expressly a matter of consultation and not of conjoint decision. But precisely because this is the situation, in this country over the last fifteen years we have succeeded in creating a genuine relation of trust between science and governmental administration, and hence these consultative groups have been able, in my view, to do very valuable work. University teachers learn there to have regard not only for the interests of the universities and research but also for the interests of society, and to act upon this responsibility. In some cases, indeed, acting from this very sense of responsibility, scientists have exerted influence on decisions in politics generally. I have in mind the declaration of the eighteen Göttingen professors on the question of arming the Bundeswehr with atomic weapons, and the fact that, some years previously, the same University of Göttingen forced the resignation of the Lower Saxon Minister of Culture, who in the University's opinion had been too closely associated with Nazi groups. I believe that this reciprocal influence, and the relation of trust that is certainly prerequisite for such influence, was a boon to the development of the Federal Republic in the postwar years, and that we should bend all our efforts to preserving this relation of trust, whatever changes of government may occur.

Finally, as a consequence of the increasing importance of science and technology, even international relations have been changed and strengthened, between universities no less than between states. International cooperation between universities in different countries, or again between national universities and international centers of research, is part of the regular picture of present-day scientific inquiry. It has become almost the rule for young scientists to pursue their research abroad for some years with the aid of stipends, so as to become acquainted with new research methods and avenues of inquiry in other countries. In the Federal Republic, the Alexander von Humboldt Foundation each year provides stipends for three hundred to four hundred young scientists from abroad, whereby they can continue their scientific work for one or two years and round out their knowledge in this country. More recently, too, the scientific attachés at the larger embassies have been seeing to it that exchange of information in the scientific and technical field shall be carried on as intensively as possible, to the benefit of both sides. Cultural policy has nowadays become a most important part of foreign policy, and its scientific and technical sector, in particular, has a significance that can hardly be overestimated.

By now, perhaps we have obtained a first general impression of the changes directly emanating from the expansion of science in the university field, changes that are having a severe effect upon the structure of the university. We must now go on to speak of those profounder alterations in the life of the university that have their origin in the changes in society, and are to that extent caused only indirectly by science and technology. For it has to be realized that all these rapidly succeeding political happenings, the revolutions, victories, defeats and conquests, however dreadful they may be in detail, are merely an accompaniment to the slower but more fundamentally operative shifts in the basis of human existence that have been brought about during the last 150 years by natural science and technology. The great mass of mankind became aware of this only with the dropping of the first atomic bomb over Japan in 1945, but of course these shifts had already begun much earlier; the catastrophe of Hiroshima was an alarm signal that was to forbid us simply to go on doing what we had done before.

At first, to be sure, the development of technology was a great success. Despite all the pessimistic forecasts of the nineteenth century, in industrial countries, it proved possible to do away almost entirely with mass impoverishment. The standard of living even of the previously poor classes rose in many cases to that of the former bourgeoisie, and distinctions of rank were wiped out. Accordingly, the contentment of these classes also seems greater nowadays than ever before. One consequence of this enlargement of the general welfare is the claim to education that is now being advanced by large numbers of people and has resulted in their thronging into colleges and universities. This claim is justified, and we should rejoice at the growing numbers of those who have learned to think critically and carefully. Even academic education is admittedly no safeguard against prejudice and ideological delusion, but it can perhaps preserve one from an excessive narrowness of thought. With these growing numbers of students, the universities have been faced with very difficult tasks, which can hardly be overcome without changes in their internal structure. But I shall reserve some brief remarks about university reform for later.

A further political consequence of technological expansion, especially in the atomic field, is the formation of large political units, which are gradually replacing the many independent nation-states of former days. For university life this process is important in that what happens at the universities in one country soon reacts upon those elsewhere, especially within the same power bloc. To the academic young, the continuance of study or research in another country of the same regional group now seems well-nigh commonplace and by no means an especially crucial step. Only the occasional migration to one of the other major regions, where the social structure is radically different, is still felt to be an adventure. The consolidation of young academics of different countries and races has already gone much further nowadays than that of the nations—a highly encouraging development but one whose political relevance should not be overestimated.

The strongest effects, however, of the expansion of science and technology are to be seen in our life style itself, in the ever-increasing dependence of the individual upon state-administered services, such as health care, river and water maintenance, public transport and roads, the control of trade and commerce. Society is endeavoring so far as possible to eliminate all risks for the individual. But perhaps too little consideration is given to the possibility that the scope of the individual in regard to human relations will then be critically reduced. We are falling into the danger of approximating that technically perfected "brave new world" so horrifyingly described, as a warning negative utopia, by Aldous Huxley.

It therefore seems to us actually a blessing that the bounds to technological expansion have now become clearly visible, and at this point new and important tasks again arise for science in the universities—and also for sociology and law. In industrial countries the constant spread of technology has so greatly altered the environment in which we have to live as to create threatening damage and danger to the population. This has lately been pointed out in many quarters, and I need expend no more words on the subject. It will certainly need great efforts to guide development here along the right paths. I venture to formulate the task confronting us in more general terms, as follows: It will be a matter of conceding to technology only so much space for expansion as will serve the real interests of human society, and of filling out that space as rationally as possible. We should, however, no longer do everything that we are technically able to do. As regards the task of the university in this connection, it would be quite absurd to insist that work should everywhere be done on problems of environmental protection alone; there must be fundamental and applied research in many different areas, because all new knowledge can do good if it is properly employed. But immediate collaboration among universities, industry and officialdom to solve this central problem of restraining technology is certaintly a most urgent task. The following analogy may be allowable here: Just as in the past two centuries even basic research has been influenced by the thought of its possible application—whether this application was regarded as a significant criterion of correctness or as a useful by-product—so scientific work in the future could receive a powerful impetus from the task of again becoming master in one's own house; in short, technology must be subordinated to real human needs.

The economic sciences also will be able to furnish important preliminary contributions here. The growth in the annual production of commodities still ranks as the most significant criterion for a healthy economy. But the time might soon come when a decline in commodity production could be better for man's welfare than an increase, and when there will have to be careful distinction made between goods which are unconditionally necessary and those others which we could just as well do without. The settlement of these burning questions is likely to be more important for the future than the only too often repeated theoretical dispute about the advantages of the various possible economic systems. The present condition of continuing economic growth is obviously not stable, and the question is only whether the braking distance is still sufficient to avoid serious catastrophes. Great efforts must be made, in any case, toward a restoration of economic equilibrium; and since this is as much a matter of the solving of numerous small questions of detail as of great fundamental decisions, the younger generation at the universities is confronted with an abundance of important tasks which it must resolve in its own most vital interests.

One may hope that from this angle it may be possible to deal with that narrowing of human relationships which seems to go along with technical advance and prosperity and which the student generation justifiably rejects. In university life this narrowing—or to put it more plainly, isolation—is evinced in the fact that although we give a young man all possible assistance in the way of grants and student housing, in order to facilitate regular study, he can still, in his overcrowded lectures and seminars, make but little contact with the professors and instructors; and even the commerce with people in other walks of life, which used formerly to go on automatically in student lodgings, is absent in the student dormitory. After a course of studies so well organized but so poor in human contacts, we cannot expect the student to escape all personal difficulties in letting himself be fitted, as a sort of spare part, into the gigantic, rationally programmed machinery of modern society—equipped, of course, with its equally preprogrammed freedoms. No younger generation has yet been ready to do this, and we ought not even to wish it. For the young, the world is always beginning anew. They cannot simply step into what the older generation has created and call it good without reservation.

Thus it is that in the past five years we have had crisis in the university, a student rebellion that has cut deeply and in part destructively into academic life. This rebellion began not in Germany but in the American West, at Berkeley, and it would therefore be wrong from the start to try to blame the crisis on German students, German professors or the German university in particular. Nor has this student rebellion been the first we have had, and it will not be the last. At the Wartburg festival of 1817 the students embraced the cause of future German unity, but three years later succumbed to the reaction that set in after the murder of the dramatist Kotzebue by the theology student Karl Sand. In 1848 the students and some of the professors demonstrated together for liberal and democratic reforms and against the continuing existence of petty princedoms. With the founding of the Bismarckian Reich, some part at least of these desires were satisfied in the abolition of petty states, but only a few of the liberal and democratic reforms were realized. The youth movement of fifty years ago sought to escape from the narrowness of a bourgeois world apparently grown hollow and untrustworthy into the wider spaces of a communal existence amid nature, a direct association with people in all walks of life, an association to be sustained by the common cultural tradition. But by the end of the twenties, large sections of the student body had gone over to National Socialism; as early as 1931 it was apparent in the rioting at Munich University that the Nazis had gained control of the student representative bodies, here as at other universities.

In the past five years, lectures have again been disrupted and elections obstructed by force. But I have not myself been a party to these events; and since I could only pledge myself to observe and describe from a distance, I may perhaps be allowed to inquire into the common features of these four student rebellions. At the start, in every case, we find a spontaneous youthful consciousness of a collective breakaway, for which it seems that no rational explanation can be given. To account for it in terms of previously existing grievances is to state a part of the truth, no doubt, but it misses essential features of the matter. Comparison with the departure of migrating birds, flying south in autumn, contains another part of the truth. As Jacob Burckhardt says: "The message runs through the air, and on the one thing that matters they are all suddenly in agreement, even were it only a dark 'things have got to change.'" At this initial stage the positive aspects of the movement predominate, the gaze is still directed outward to the new and alluring possibilities, and even the confusions of this early phase demonstrate that the struggle is still for true values and not merely for power.

But it cannot stay that way, and there follows a period in which the old political forces endeavor to guide the waters of this new spring into the perhaps already rather sluggish and turbid stream of their own political aims. After 1848, it was Bismarck's Germany that eventually held the field. But in the twenties, it was the long-standing historical movements of nationalism and socialism whose unholy alliance in National Socialism bemused the minds of the young. Anyone who lived through that time can only remember with horror how, even among highly gifted young men with the wide world at their feet, the viewpoint suddenly narrowed, how it seemed to be rigidly fixed upon a mere parcel of obvious grievances whose removal was thought to promise deliverance from all evil. Prophets, who always appear at such times, then fabricate a new language, which renders an understanding with those not gripped by the movement more difficult than ever, and hence the chance of some good emerging from the first outbreak grows increasingly small. It is just this constriction of the field of vision, this ideological blinkering, that can lead young men to follow those who no longer speak truth, who demonstrate for peace because they are arming for civil war, who talk of freedom and mean the repression of conscience. But a struggle conducted in blindness can hardly serve a reasonable purpose. Such a thing should not happen again, and even in the present crisis we must make every effort to prevent this contraction of the field of view. Only one example is needed to show that this danger is still with us today. It is easy to see that science can be misused, e.g., in weapons technology. But perhaps it is not recognized that there are also dangerous misuses of psychology. The most dangerous consists in a primary imputation of dishonest motives to the actions of those who think differently, whereby mistrust and enmity are the inevitable consequences.

But how far can we acquiesce in the demands of the students? This is a question I have no right to answer, since I no longer take an active part in university life. But perhaps I may still offer a piece of advice. So far as the desires of the young are really directed to the improvement of conditions in the universities—improvement in regard to the goals that are set for the universities by society—we should try, in a common effort with them and in complete understanding of their considerable human problems, to improve things where improvements can be made; we should not, by pointing out many erroneous courses, allow ourselves to be misled into just sluggishly trailing on in the old ways. But where the young are trying to effect radical changes in society from within the university, we should make it clear to them that the university may well be an unsuitable place for this. The university plays a much smaller role in political life than we would like to imagine. To the mass of men, action has always been more important than thought.

But now a few more words on university reform, which is so much discussed at present. It can be no business of mine to make proposals on that subject. Although an adaptation of the university to changing times is indispensably necessary, it is really no longer a matter of making proposals but rather of carrying them out. And for this purpose one must be either in the midst of academic life or on the margin between that and public life. As for the constant new proposals, it is perhaps worth remembering that in a Rectoral Address of 1899 there was already talk of the "feeling of weariness that overtakes one on repeatedly seeing some university teacher step up to the Danaïds' tub of academic reform proposals, after so many excellent men have been doing it for generations." So there will be no new proposals from me. Of the urgent need to shorten the curriculum I have already spoken earlier. Beyond that I shall dwell particularly on just one further point, which has to do with the concept of democracy. The dispute about specific percentages in joint decision-making strikes me as resembling the struggle of children over a toy that they have long since broken in the quarrel, and in which it can no longer matter how large a piece each of them retains. The one thing that matters in university life is the restoration of a bond of trust between professors and students. If such a relationship exists, good collaboration will be possible whatever the percentage; if it does not, the work cannot yield much in any event. Democracy, we may be sure, is not just a set of political rules of play; it is a way of life which starts by conceding full status to the other party, which takes him seriously as a person and tries to find a solution with him, not against him.

But I have strayed rather far from my topic of "Science in the Contemporary University," and would like to return to the question whether the scientific mode of thought, which has so visibly gained ground in the university, is compatible with the desire for that breadth and freedom of thought that have always seemed to us the most important goals of academic education.

Let us begin by recalling that at the outset of modern times it was natural science that brought liberation from a narrow, dogmatic way of thinking handed down from the Middle Ages. Later on, owing to the demand for care and exactness in detail, the view has again also narrowed at times. But if the scientist is charged with an overspecialized, professional mode of thought, it can be pointed out that the picture has changed here during the last fifty years; that in present-day science the narrow and conscientious specialist, though he still has an important place, can no longer play the leading role. For whether it be in physics, chemistry, biology or medicine, we are compelled to look out across the frontiers into neighboring territory, and often beyond this territory into philosophy itself, if we wish to make fundamentally important advances and to understand them. Above and beyond this there is the ethical and educative relevance of scientific research. The practical significance of our research findings compels us to reflect anew upon ethical problems, more especially the question whether one should do everything that one is able to do. The precision of scientific thinking teaches us that truth and falsity are ultimately decided objectively, that here subjective opinion and personal commitment may indeed be important for the work but that they do not suffice to make us right, and that we may also turn out, on the contrary, to be wrong. That is an exceedingly valuable thing to learn. Finally, in atomic physics and biology, science encounters the epistemological boundaries of rational thought, which it attempts to mark out by means of the rational expedients of knowledge. Thus even here the student is confronted with all the currents of thought whose acquaintance was called for in Wilhelm von Humboldt's ideal of education. Breadth of thought is a crucial precondition even in natural science.

But admittedly, that is not yet enough for the young; from their encounter with the university they hope to get answers to their questions about life, a compass by which to orient themselves in the attempt to give their own lives a meaning. As an institution, the university is certainly no better able to provide this nowadays than it was before. But even today, the young student may still perhaps be lucky enough to encounter at the university some particular significant personality who helps him on, and such an encounter can affect his whole life, regardless of what field of learning it occurs in. At the University of Munich such personalities have always been found. But in honesty we must tell the young man that he cannot safely rely on this. In the course of his studies perhaps he will also make friends in whose company he will more easily find his way into life; but otherwise he must content himself with the bitter truth that in the most important decisions we are always ultimately alone.

With this we arrive at the problem of the value of freedom. Freedom, of course, invariably has two aspects, freedom from and freedom to. In the case of mental freedom, it is a matter, on the one hand, of freedom from prejudices, from dogmatic ties, from suggestive influences, from an imposed point of view; and on the other of being able to think new thoughts, to look at known facts with fresh eyes, to follow the thoughts of others, even if they do not at first illuminate, and to go beyond them. From science one may learn above all things here that freedom is possible only through acknowledgment of laws. The doctor can free the patient from his sufferings only if he knows and utilizes the biological laws that govern the workings of the organism. The freedom to fly depends upon an acknowledgment of the laws of aerodynamics. And freedom in the decisions of life is likewise possible only through adherence to moral norms, and anyone who thinks to disdain these as a form of coercion would be merely replacing freedom by lack of principle. For science also teaches something else of great importance, namely that freedom is difficult. To perceive new connections amid the inexorable laws of nature, to explore new possibilities, to think in unaccustomed ways, can be achieved only by the utmost effort. But anyone who finds it too difficult should not be led astray into simply ignoring the existing laws. Nothing whatever would come of that. He will then be well advised, however, to stay within the framework of what already exists, and to carry out his work with care; that is always worth while. Wilhelm von Humboldt may have called upon the university to train men in critical, scientific thinking, to instill in them a conscious moral will directed to a grasp of principles; even so, the expansion of natural science certainly offers no obstacle to this goal.

And yet there is a great deal of dissatisfaction and unrest in the university today. To many of those who call for more freedom and democracy, it may be that freedom and democracy are in reality too difficult. Democracy ought not to be so, at bottom; for the requirement of taking the other seriously as a person should be capable of fulfillment with goodwill, even when the other stands in our way to obstruct us. But in our own day, when many old values are called in question, when truth and untruth are inextricably mingled in a Babel-like confusion of tongues, it is actually very difficult to find one's way here alone, to decide for oneself where the ground is still firm underfoot and where it is beginning to quake; we should not hold it against the young if they are not willing to bear the burden of freedom. But if we are honest, we should then advise them to cling to the old standards of value, which have been conserved in the great religions; for the time has not yet come for the writing of a new canon. A rationalistic analysis of social relationships is certainly not adequate for that.

The human relation between the old and the young also requires that we should not treat the irrational features of contemporary youth movements, no less than earlier ones, as a reason for abandoning understanding. The dominating influence of science and technology has so overstressed the rational aspect of the world that a reaction against this overemphasis seems quite unavoidable; or, to put it in Nietzsche's words, that in desperation at the emptiness and suffering of such a world the god Dionysus should again make his appearance. In all these irrational doings there is probably an unconscious expression of longing for that world in which mind is more than information, love more than sexuality and science more than the collection and analysis of empirical data. Let us therefore be thankful that life itself is constantly giving rise to movements that have not been preprogrammed in Huxley's brave new world. Within this situation, however, the university is faced with the crucially important task of preserving breadth of vision in the young, of making their thinking mobile enough to escape ossification in cheap dogmatic forms, of allowing them to share in the great possibilities that our time of transition has to offer. That can be done in science no less than in other fields of study, in concentration on a particular problem no less than in comprehensive surveys over a broad field.

For five centuries the University of Munich has proved a source and storehouse of fruitful thought; it will also be able to face the tasks set by our present age.

XVI

Scientific and Religious Truth

The honor you have so kindly accorded me, and for which I thank you, is linked with the name of Romano Guardini. It has thereby acquired a special value to me, for even at an early age the spiritual world of Guardini left a deep impression on me. As a young man I read his writings and saw the characters from Dostoevski's novels as he illuminated them, and in later life I had the pleasure of knowing him personally. This world of Guardini's is through and through a religious and Christian one, and at first it seems hard to establish any link between it and the world of the sciences, in which I have worked since my student days. In the history of science, ever since the famous trial of Galileo, it has repeatedly been claimed that scientific truth cannot be reconciled with the religious interpretation of the world. Although I am now convinced that scientific truth is unassailable in its own field, I have never found it possible to dismiss the content of religious thinking as simply part of an outmoded phase in the consciousness of mankind, a part we shall have to give up from now on. Thus in the course of my life I have repeatedly been compelled to ponder on the relationship of these two regions of thought, for I have never been able to doubt the reality of that to which they point. In what follows, then, we shall first of all deal with the unassailability and value of scientific truth, and then with the much wider field of reli-

Speech before the Catholic Academy of Bavaria, on acceptance of the Guardini Prize. *Frankfurter Allgemeine Zeitung,* March 24, 1973, pp. 7–8.

gion, of which—so far as the Christian religion is concerned—Guardini himself has so persuasively written; finally—and this will be the hardest part to formulate—we shall speak of the relationship of the two truths.

Of the beginnings of modern science, the discoveries of Copernicus, Galileo, Kepler and Newton, it is usually said that the truth of religious revelation, laid down in the Bible and the writings of the Church Fathers and dominant in the thought of the Middle Ages, was at that time supplemented by the reality of sensory experience, which could be checked by anyone in possession of his normal five senses, and which—if enough care was taken—could therefore not in the end be doubted. But even this first approach to a description of the new way of thought is only half correct; it neglects decisive features without which its power cannot be understood. It is certainly no accident that the beginnings of modern science were associated with a turning away from Aristotle and a reversion to Plato. Even in antiquity, Aristotle, as an empiricist, had raised the objection—I cite more or less his own words—that the Pythagoreans (among whom Plato must be included) did not seek for explanations and theories to suit the facts, but distorted the facts to fit certain theories and favored opinions, and set themselves up, one might say, as coarrangers of the universe. In fact the new science led away from immediate experience in the manner criticized by Aristotle. Let us consider the understanding of the planetary motions. Immediate experience teaches that the earth is at rest and that the sun goes around it. In the more precise terms of our own day, we might even say that the word "rest" is defined by the statement that the earth is at rest, and that we call every body at rest that no longer moves relative to the earth. If the word "rest" is understood in this fashion—and it generally *is* so understood—then Ptolemy was right and Copernicus wrong. Only if we reflect upon the concepts of "motion" and "rest," and realize that motion implies a statement about the relation between at least two bodies, can we reverse the relationship, making the sun the still center of the planetary system, and thereby obtaining a far simpler and more unified picture of the solar system, whose explanatory power was later fully recognized by Newton. Copernicus has thus appended to immediate experience a wholly new element, which I shall describe at this point as the "simplicity of natural laws," and which in any event has nothing to do with immediate experience. The same can be seen in Galileo's laws of falling bodies. Immediate experience teaches that light bodies fall more slowly than heavy ones. Galileo maintained, on the contrary, that in a vacuum all bodies fall with equal speed, and that their fall could be correctly described by mathematically formulable laws, namely the Galilean laws of falling bodies. But motion in a vacuum was at that time still quite impossible to observe. The place of immediate experience has therefore been taken by an idealization of experience, which claims to be recognized as the correct idealization by virtue of the fact that it allows mathematical structures to become visible in the phenomena. There can be no doubt that in this early phase of modern science the newly discovered conformity to mathematical law has become the true basis for its persuasive power. These mathematical laws, so we read in Kepler, are the visible expression of the divine will, and Kepler breaks into enthusiasm at the fact that he has been the first here to recognize the beauty of God's works. Thus the new way of thinking assuredly had nothing to do with any turn away from religion. If the new discoveries did in fact contradict the teachings of the Church at certain points, this could have little significance, seeing that it was possible to perceive with such immediacy the workings of God in nature.

The God here referred to is, however, an ordering God, of whom we do not at once know whether He is identical with the God to whom we turn in trouble, and to whom we can relate our life. It may therefore be said, perhaps, that here attention was directed entirely to one aspect of the divine activity, and that hence there arose the danger of losing sight of the totality, the interconnected unity of the whole. But this too, once more, was precisely the reason for the abundant fruitfulness of the new natural science. So much had already been said about the larger scheme of things by philosophers and theologians that there was no longer much new to say about it; scholasticism had produced weariness of thought. But the details of natural processes had as yet been scarcely looked into. This was also work in which many lesser spirits could take part, and besides, the knowledge of these details promised practical benefits. In some of the scientific societies that then came into being, it therefore became an absolute principle that only observed details should be discussed, not the larger connection of the whole. The fact that the concern was with idealized rather than immediate experience led to the development of a new art of experimentation and measurement, whereby the attempt was made to approximate the ideal conditions, and it turned out that agreement can always be reached about the results of experiment. That is not indeed so self-evident a matter as it has appeared to later centuries; for it presupposes that under the same conditions the same thing always happens. It was found, then, that if particular phenomena are defined and isolated from the environment by carefully chosen experimental conditions, it emerges clearly that they are subject to law, and are governed by a uniform chain of causal connection. Confidence in the causal succession of events, which were taken to be objective and independent of the observer, was thereby erected into a basic postulate of modern natural science. This postulate has proved itself admirably for centuries, and only in our own day have we been informed, by discoveries in the atomic field, of the limits set to this procedure. But even if these discoveries are taken into account, we have gained a seemingly unassailable criterion of truth. The repeatability of experiments always makes it eventually possible to agree about the true behavior of nature.

This general tendency of the new science also foreshadows a characteristic feature that has often been discussed, namely the emphasis on the quantitative. The demand for precise experimental conditions, accurate measurements, an exact, unambiguous terminology and a mathematical presentation of the idealized phenomena has determined the aspect of this science of nature, and brought it the name of "exact science." This term is sometimes uttered in commendation and sometimes in reproach. In commendation, when emphasis is laid upon the reliability, exactitude and unassailability of its statements; in reproach, when the intention is to intimate that such a science is unable to do justice to the infinite wealth of qualitatively distinct experiences, that it is too narrow in scope. In our own day this aspect of natural science, and of the technology that springs from it, has emerged much more sharply than ever before. We have only to think of the extreme demands for precision required in a moon landing, the almost unimaginable degree of reliability and exactitude demonstrated there, in order to recognize how firm is the footing upon which modern science bases its claim to truth.

But now we naturally have also to ask how valuable are the achievements that can be attained by this concentration on one partial aspect, this confinement to a single special portion of reality. To this question, our age gives a divided answer. We speak of the ambivalence of science. We have found that in those parts of the world where this combination of science and technology has prevailed, the material misery of the poorer sections of the population has largely disappeared, that modern medicine is preventing the massive loss of life through epidemic diseases, that transportation and communication techniques are making every day life easier. On the other hand, science can be misused to develop weapons of the most fearful destructiveness; the rapid growth of technology impairs and threatens our environment. Even apart from these immediate dangers, standards of value are shifting; attention is too much drawn to the narrow field of material welfare, and the other foundations of our existence are neglected. Even if technology and science could be employed merely as means to an end, the outcome depends upon whether the goals for whose attainment they are to be used are good ones. But the decision upon goals cannot be made within science and technology; it is made, if we are not to go wholly astray, at a point where our vision is directed upon the whole man and the whole of his reality, not merely on a small segment of this. But this total reality contains much of which we have not said anything yet.

First there is the fact that man can develop his mental and spiritual powers only in relation to a human society. The very capacities that distinguish him above all other living creatures, the ability to reach beyond the immediate sensory given, the recognition of wider interrelations, depend upon his being lodged in a community of speaking and thinking beings. History teaches that such communities have acquired in their development not only an outward but also a spiritual pattern. And in the spiritual patterns known to us, the relation to a meaningful connection of the whole, beyond what can be immediately seen and experienced, has almost always played the deciding role. It is only within this spiritual pattern, of the ethos prevailing in the community, that man acquires the points of view whereby he can also shape his own conduct wherever it involves more than a mere reaction to external situations; it is here that the question about values is first decided. Not only ethics, however, but the whole cultural life of the community is governed by this spiritual pattern. Only within its sphere does the close connection first become visible between the good, the beautiful and the true, and here only does it first become possible to speak of life having a meaning for the individual. This spiritual pattern we call the religion of the community. The word "religion" is thereby endowed with a rather more general meaning than is customary. It is intended to cover the spiritual content of many cultures and different periods, even in places where the very idea of God is absent. Only in the communal modes of thought pursued in modern totalitarian states, in which the transcendent is completely excluded, would it be possible to doubt whether the concept of religion can still be meaningfully applied.

How strongly the stamp of religion is impressed on the visage of a human community, and on the life of its individual members, can scarcely be better described than it is by Guardini in his book on the characters in Dostoevski's novels. The life of these characters is filled at every moment by the struggle for religious truth and is somehow imbued with the spirit of Christianity, and hence it does not matter so very much whether these people are victorious or defeated in the struggle for goodness. Even the greatest villains among them still know what is good and what bad, measuring their deeds by the guiding ideals that the Christian faith has given them. Here also there lapses the familiar objection to the Christian religion, that men in the Christian world have behaved just as dreadfully as those outside it. That is unfortunately true, no doubt, but these men preserve in themselves a clear power of distinguishing good from bad; and only where this is still present does the hope of improvement remain. Where no guiding ideals are left to point the way, the scale of values disappears and with it the meaning of our deeds and sufferings, and at the end can lie only negation and despair. Religion is therefore the foundation of ethics, and ethics the presupposition of life. For every day we must meet decisions, and must know or at least have an idea of the values whereby we govern our conduct.

At this point we also recognize the characteristic difference between genuine religions, in which the spiritual realm, the central spiritual order of things, plays a decisive part, and the narrower forms of thought, especially in our own day, which relate only to the strictly experienceable pattern of a human community. Such forms of thought exist in the liberal democracies of the West no less than in the totalitarian states of the East. Here, too, to be sure, an ethic is formulated, but the talk is of a norm of ethical behavior, and this norm is derived from a world outlook, that is, from inspection of the immediately visible world of experience. Religion proper speaks not of norms, however, but of guiding ideals, by which we should govern our conduct and which we can at best only approximate. These ideals do not spring from inspection of the immediately visible world but from the region of the structures lying behind it, which Plato spoke of as the world of Ideas, and concerning which we are told in the Bible, "God is a spirit."

Religion, however, is not only the foundation of ethics; it is also above all—and this, too, we can learn from Guardini—the foundation of trust. Just as we learn language as children, and feel the understanding made possible in language to be the most important constituent of trust in men, so the images and likenesses of religion, which themselves represent a kind of poetic language, produce trust in the world, in the meaning of our existence. The fact that there are many different languages is no sort of objection, any more than is the fact that we are born seemingly by chance into a particular linguistic area or religious territory, whose marks we bear. What matters is only that we are led on toward the world in this trust, and that can happen in every language. For the native Russians, for example, who appear in Dostoevski's novels and of whom Guardini writes, the workings of God in the world are a constantly repeated immediate experience, and thus their trust is renewed again and again, even when outward privation seems inexorably to stand in its way.

Finally, religion, as I have said, is of decisive importance for art. If by religion we designate, as we have done, simply the spiritual pattern into which a human community has evolved, it is virtually self-evident that art, too, must be an expression of religion. A glance into the history of the most diverse cultures teaches us that we can infer most directly the spiritual pattern of an earlier period from the works of art that survive, even when the religious doctrine in which the spiritual pattern was formulated is scarcely known to us.

But all that has here been said about religion is naturally well known to you as members of this group. It has been repeated only in order to emphasize that even the natural scientist must recognize this comprehensive significance of religion in human society, if he wants to try to think about the relation of religious and scientific truth. The fact that these two truths have come into conflict has exercised a decisive influence on the history of

European thought since the seventeenth century. The starting point of the conflict is generally said to have been the prosecution of Galileo by the Roman Inquisition in 1616, where the question at issue was the teaching of Copernicus, the quincentenary of whose birth was celebrated recently.

We must now go into this starting point in a little more detail. Galileo had endorsed the theory of Copernicus, whereby —in contrast to the then generally prevailing Ptolemaic view of the world—the sun is at rest in the center of the planetary system, while the earth revolves about the sun and also rotates on its own axis in twenty-four hours. Galileo's student B. Castelli had declared that the theologians must now explain the Bible in accordance with the established facts of natural science. Such a statement could be seen as an attack on holy scripture, and the Dominican fathers Caccini and Lorini brought the matter before the Roman Inquisition. In the judgment of February 23, 1616, the two statements attributed to Copernicus in the indictment—"The sun is the center of the universe and therefore immovable," and "The earth is not the center of the universe and immovable, but moves daily, even about itself"— were declared to be philosophically absurd and heretical. With the consent of Pope Paul V, Cardinal Bellarmine was commissioned to warn Galileo to abandon the Copernican theory. If he refused, the Cardinal was to order him not to teach, defend or discuss such a view. Galileo complied with this order with some loyalty, but especially after Urban VIII had ascended the papal throne Galileo believed that he was again allowed to prosecute his inquiries openly. After the publication of his celebrated polemical treatise, the *Dialogo,* in 1632, a second trial resulted, at which Galileo was obliged to forswear Copernicanism in all its forms. The details of the trial need no longer engage our interest nowadays—not even the human shortcomings which played a part on both sides. But we can and must inquire into the deeper grounds of the conflict.

It is important to realize, in the first place, that both parties must have believed themselves in the right. The ecclesiastical authorities and Galileo were both convinced that major values were in peril, and that it was their duty to defend them. Galileo had found—as I mentioned at the outset—that on careful observation of terrestrial and celestial phenomena, in the fall of a stone no less than in the motions of the planets, mathematical regularities come to light that permit a hitherto unheard-of degree of simplicity to become visible in the phenomena. He had perceived that from this simplicity there emerges a new possibility of understanding, that here we can trace out in our thinking segments of the eternal order of the phenomenal world. The Copernican interpretation of the planetary system was simpler than the traditional Ptolemaic one; it provided a new kind of understanding, and Galileo was unwilling on any terms to be deprived of this new insight into the divine order. The Church, on the other hand, believed that there should be no disturbance of the view of the world that for many centuries had been a self-evident part of Christian thinking, if no wholly compelling reasons were present. Such compelling reasons neither Copernicus nor Galileo was able to produce. As a matter of fact, the first proposition of the Copernican theory under discussion here was quite certainly false. Science today would equally decline to say that the sun is at the center of the universe, and therefore immovable. As to the second proposition, referring to the earth, it would first be necessary to elucidate the meaning of the words "rest" and "motion." If an absolute meaning is ascribed to them, as is done by naïve thought, it is simply a matter of definition that the earth is at rest, or at any rate that we use the word thus and not otherwise. Once it is seen that the concepts have no absolute meaning and that they refer to the relation between two bodies, it becomes arbitrary whether the sun or the earth is regarded as in motion or at rest. In that case, there is really no reason to change the old picture of the world.

It may be suspected, nonetheless, that the members of the Inquisition tribunal were fully aware of the power implicit in the concept of simplicity which was upheld here—consciously or otherwise—by Galileo and which was connected at the philosophic level with the reversion from Aristotle to Plato. The judges also obviously had the greatest respect for Galileo's scientific authority; hence they did not want to hinder him in the prosecution of his researches, but were anxious to avoid creating disturbance and uncertainty in the traditional Christian concept of the world, which had played such a crucial part in the structure of medieval society and continued to do so. Scientific findings, after all, especially when they are novel, are often subject to some degree of change; the final judgment can generally be passed only after some decades of testing. Why should not Galileo wait before publication? It will thus have to be granted to the Inquisition tribunal that at the first trial they were trying to reach a settlement and arrived at a justifiable decision. It was only later, when Galileo had exceeded the bounds imposed by Bellarmine's admonition, that the upper hand could be gained at the second trial by those to whom force seemed simpler than the attempt at compromise, and hence there followed that notoriously severe judgment against Galileo which has subsequently been so damaging to the Church.

What weight would we nowadays attach to the argument that one should not prematurely bring unrest and uncertainty into a view of the world which, as a component of the spiritual structure of society, plays an important part in harmonizing the life of the community? To this argument many radical spirits would nowadays react with scorn; they would point out that what is at issue here is merely the preservation of obsolete power structures, and that hence, on the contrary, we should see to it that such social structures are changed or abolished as rapidly as possible. But it is necessary to remind these radical spirits that the conflict between science and the prevailing view of the world still goes on at the present day, and does so, indeed, in the totalitarian states where dialectical materialism has been adopted as the basis of thought. Thus official Soviet philosophy has found it difficult to accommodate itself to relativity theory and the quantum theory; on questions of cosmology in particular, there have been violent clashes of opinion there. In 1948 a congress on ideological questions in astronomy was finally held in Leningrad, which was intended to clarify the disputed problems by discussion and agreement and has contributed to settling them.

As in the trial of Galileo, the fundamental issue here is by no means concerned with questions of fact but rather with the conflict between the spiritual pattern of a community, which by nature has to be a static thing, and the constantly expanding and novel findings and modes of thought in science, which is thus a dynamic affair. Even a society that has emerged from great revolutionary upheavals endeavors to consolidate and fix the intellectual heritage that is to form the enduring basis of the new community. Complete uncertainty about all standards would be intolerable in the long run. Science, however, strives to extend itself. Even if natural science, or some other kind of science, is to be used as the basis for a world outlook—and something of the sort is attempted in dialectical materialism—it can only be the science of past decades or past centuries, and once this has been fixated in language, the preconditions for a later conflict have again been created. It therefore seems better to make it clear at the outset, through the images and likenesses for the larger scheme of things, that here we are speaking a poetic language, a language open to all human values and replete with living symbols, not a language of science.

Despite these general difficulties, we must return once more to the factual issues in the prosecution of Galileo. Was it of importance to the Christian community that Copernicus should construe certain astronomical observations in a different fashion from Ptolemy? For the Christian conduct of individual life it might basically have been a matter of complete indifference whether there are or are not crystalline spheres in the heavens, whether the planet Jupiter is girdled with moons, whether the earth or the sun is situated at the center of the universe. For such an individual, after all, the earth was in any case at the center; it was his dwelling place. But for all that, it was not a matter of indifference. Two hundred years later, Goethe could still express his dread and admiration at the sacrifices that had had to be made in accepting the Copernican theory. He himself made them only reluctantly, even though persuaded that the theory was correct. Perhaps the same doubts had arisen already, consciously or otherwise, among the judges of the Roman Inquisition over whether the science of Galileo might not bring about a dangerous shift of perspective. And yet even they were unable to deny that when the scientist, like Galileo or Kepler, discovers mathematical structures in the phenomena, he thereby makes visible partial uniformities in the divine ordering of the universe.

But this very vision of the dazzling parts might perhaps obscure the vision of the whole; it might have as its consequence, insofar as the connection of the whole is eclipsed in the consciousness of the individual, that the living coherence of the human community also suffers and is threatened with decay. With the replacement of the natural conditions of life by technically contrived processes, there also enters an estrangement between the individual and the community that produces dangerous instabilities. In his play *Galileo,* Bertolt Brecht has a monk say, "The decree against Copernicus has revealed to me the dangers that all-too-untrammeled inquiry harbors for mankind." We do not know whether this motive was already operative at that time, but we have learned in the meantime how great the dangers are.

We have learned still more from the development of natural science in a European world that bears the impress of the Christian religion, and this will be the topic of the last part of my address. I have already sought to enunciate the thesis that in the images and likenesses of religion, we are dealing with a sort of language that makes possible an understanding of that interconnection of the world which can be traced behind the phenomena and without which we could have no ethics or scale of values. This language is in principle replaceable, like any other; in other parts of the world there are and have been other languages that provide for the same understanding. But we are born into a particular linguistic area. This language is closer akin to that of poetry than to the precision-orientated language of natural science. Hence the words in the two languages often have different meanings. The heavens referred to in the Bible have little to do with the heavens into which we send up aircraft and rockets. In the astronomical universe, the earth is only a minute grain of dust in one of the countless galactic systems, but for us it is the center of the universe—it really is the center. Science tries to give its concepts an objective meaning. But religious language must avoid this very cleavage of the world into its objective and its subjective sides; for who would dare claim the objective side to be more real than the subjective? Thus we ought not to intermingle the two languages; we should think more subtly than we have hitherto been accustomed to do.

Moreover, the development of science in the last hundred years has compelled this more subtle thinking even in its own field. Since we have ceased to make the world of immediate experience into an object of research, and have looked instead into a world we can penetrate only with the instruments of modern technology, the language of daily life is no longer adequate here. We are at last succeeding, as it happens, in understanding this world, through the representation of its patterns of order in mathematical forms; but if we wish to talk about it, we have to be content with images and likenesses, almost as we do in religious language. We have thus learned to think more circumspectly about language, and have come to realize that apparent contradictions may have their source in the inadequacy of language itself. Modern science has brought to light regularities of a very far-reaching kind, vastly more comprehensive than those which preoccupied Kepler and Galileo. But in so doing it has turned out that as the extent of the connections increases, so also does the degree of abstractness, and with it, too, the difficulty of understanding. Even the demand for objectivity, which has long ranked as the precondition for all science, has undergone restriction in atomic physics, in virtue of the fact that a complete separation of the observer from the phenomenon to be observed is no longer possible. How, then, does it stand with the opposition between scientific and religious truth?

The physicist Wolfgang Pauli once spoke of two limiting conceptions, both of which have been extraordinarily fruitful in the history of human thought, although no genuine reality corresponds to them. At one extreme is the idea of an objective world, pursuing its regular course in space and time, independently of any kind of observing subject; this has been the guiding image of modern science. At the other extreme is the idea of a subject, mystically experiencing the unity of the world and no longer confronted by an object or by any objective world; this has been the guiding image of Asian mysticism. Our thinking moves somewhere in the middle, between these two limiting conceptions; we should maintain the tension resulting from these opposites.

The care to be taken in keeping the two languages, religious and scientific, apart from one another, should also include an avoidance of any weakening of their content by blending them. The correctness of tested scientific results cannot rationally be cast in doubt by religious thinking, and conversely, the ethical demands stemming from the heart of religious thinking ought not to be weakened by all too rational arguments from the field of science. There can be no doubt, in this connection, that through the enlargement of technical possibilities new ethical problems have also appeared that cannot be easily resolved. I may mention as examples the problem of the researcher's responsibility for the practical application of his discoveries, or the still more difficult question from the field of modern medicine of how long a doctor should or may prolong the life of a dying patient. Consideration of such problems has nothing to do with any watering down of ethical principles. Nor am I able to conceive that such questions are capable of being answered by pragmatic considerations of expediency alone. On the contrary, here too it will be necessary to take into account the connection of the whole—the source of ethical principles in that basic human attitude which is expressed in the language of religion.

Today, moreover, we may already be able to effect a more correct distribution of the emphases that have been misplaced by the enormous expansion of science and technology in the past hundred years. I mean the emphases we ascribe to the material and the spiritual preconditions in the human community. The material conditions are important, and it was the duty of society to eliminate the material privation of large sections of the population, once technology and science had made it possible to do so. But now that this has been done, much unhappiness remains, and we have come to see how compellingly the individual also has need, in his self-consciousness or self-understanding, for the protection the spiritual pattern of a community can provide. It is here, perhaps, that our most important tasks now lie. If there is much unhappiness among today's student body, the reason is not material hardship but the lack of trust that makes it too difficult for the individual to give his life a meaning. We must try to overcome the isolation which threatens the individual in a world dominated by technical expediency. Theoretical deliberations about questions of psychology or social structure will avail us little here, so long as we do not succeed in finding a way back, by direct action, to a natural balance between the spiritual and material conditions of life. It will be a matter of reanimating in daily life the values grounded in the spiritual pattern of the community, of endowing them with such brilliance that the life of the individual is again automatically directed toward them.

But it is not my business to talk about society, for we were supposed to be discussing the relationship of scientific and religious truth. In the past hundred years, science has made very great advances. The wider regions of life, of which we speak in the language of our religion, may thereby have been neglected. We do not know whether we shall succeed in once more expressing the spiritual form of our future communities in the old religious language. A rationalistic play with words and concepts is of little assistance here; the most important preconditions are honesty and directness. But since ethics is the basis for the communal life of men, and ethics can only be derived from that fundamental human attitude which I have called the spiritual pattern of the community, we must bend all our efforts to reuniting ourselves, along with the younger generation, in a common human outlook. I am convinced that we can succeed in this if again we find the right balance between the two kinds of truth.

About the Author

Werner Heisenberg was born in Würzburg, Germany, in 1901. He was educated at the Universities of Munich and Göttingen and in 1932 was awarded the Nobel Prize for his work in theoretical atomic physics. He is now Director of the Max Planck Institute for Physics and Astrophysics in Munich.

About the Editor of This Series

Ruth Nanda Anshen, philosopher and editor, plans and edits *World Perspectives, Religious Perspectives, Credo Perspectives, Perspectives in Humanism* and *The Science of Culture Series.* She also writes and lectures on the relationship of knowledge to the nature and meaning of man and existence. She is the author of the recently published *The Reality of the Devil: Evil in Man,* a study in the phenomenology of Evil, published by Harper & Row.